Y0-BYB-275

ORGANIC PHOTOCHEMICAL SYNTHESES

VOLUME 2

ORGANIC PHOTOCHEMICAL SYNTHESES

VOLUME 2

1976

EDITOR

R. SRINIVASAN

IBM WATSON RESEARCH CENTER
YORKTOWN HEIGHTS, NEW YORK

ASSOCIATE EDITORS

T. D. ROBERTS

UNIVERSITY OF ARKANSAS
FAYETTEVILLE

JAN CORNELISSE

RIJKSUNIVERSITEIT
LEIDEN, THE NETHERLANDS

A Wiley-Interscience Publication
JOHN WILEY & SONS
New York · London · Sydney · Toronto

Library of Congress Catalog Card Number: 78-161147

ISBN 0-471-81921-2

Printed in the United States of America

10 9 8 7 6 5 4 3 2 1

CONTRIBUTORS

Günter Adam
W. Adam
R. J. Balchunis
Ronald S. Beckley
Kuldip S. Bhandari
Jordan J. Bloomfield
J. Brussee
R. M. Bowman
T. R. Chamberlain
A. Couture
Stanley J. Cristol
Michael Cunningham
Randall J. Daughenbaugh
David A. Dickinson
F. C. De Schryver
M. B. DeTar
Wendell L. Dilling
D. L. Dolce
Dietrich Döpp
Aubry E. Dupuy, Jr.
W. Eberbach
Rudolf E. Gayler
P. Gilgen
Joseph M. Grindel
Erling Grovenstein, Jr
Gary L. Grunewald
Thomas A. Hardy
Harold Hart
C. W. Huang
H. J. C. Jacobs
Tappey H. Jones
George Just
G. Kaupp
K. Kees
R. M. Kellogg
Paul J. Kropp
W. H. Laarhoven
Russell A. LaBar
A. Lablache-Combier
L. Leenders
L. M. Leichter
J.-C. Liu
R. S. H. Liu
H. Loos
S. Masamune
J. J. McCullough
R. C. Miller
R. D. Miller
N. Nakatsuka
M. Nastasi
D. C. Neckers
H. Ofenberg
Dennis C. Owsley
Albert Padwa
Leo A. Paquette
W. L. Prins
H. Prinzbach
V. Ramamurthy
Durvasula V. Rao
O. Rodriguez
W. H. F. Sasse
A. P. Schaap
John R. Scheffer
E. S. Schilling
H. Schmid
B. Schoustra
Klaus Schreiber
A. G. Schultz
Louis H. Stekoll
J. Strating
J. Streith
G. Subrahmanyam
J. S. Swenton
Erach R. Talaty
James W. Taylor
A. L. Thayer
A. H. A. Tinnemans
P. Uebelhart
S. I. Wetmore, Jr.
J. H. Wieringa
J. R. Williams
H. Wynberg
H. Ziffer

PREFACE

The warm reception that was accorded to the first volume in this series underscored the need to compile and publish further collections of useful synthetic procedures in organic photochemistry. The present volume contains 41 examples of such syntheses. In the years since the publication of the first volume there has been considerable progress in the discovery and investigation of new photochemical reactions. Many of them are exemplified in this volume.

We hope that workers in this field will continue to keep us informed of novel, synthetically useful organic photochemical reactions.

R. Srinivasan
T. D. Roberts
Jan Cornelisse

Yorktown Heights, New York
Fayetteville, Arkansas
Leiden, The Netherlands
February 1976

CONTENTS

ORGANIC PHOTOCHEMICAL SYNTHESES

VOLUME 2

1-ACETYL-4,5-DICARBOMETHOXY-1*H*-AZEPINE

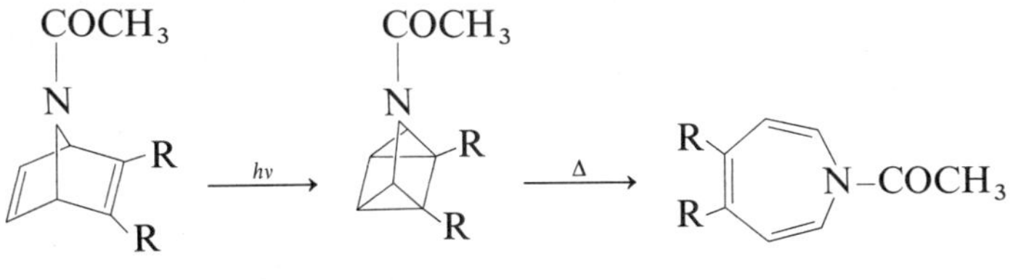

$R = CO_2CH_3$

Submitted by G. KAUPP and H. PRINZBACH*
Checked by J. CORNELISSE and Mrs. G. M. GORTER-LAROY†

1. Procedure

A solution of 7-acetyl-7-azadimethylbicyclo[2.2.1]hepta-2,5-diene-2,3-dicarboxylate (1.0 g, 0.004 mole) (Note 1) in 400 ml of absolute diethylether is cooled to −40°C in a cylindrical immersion apparatus equipped with a nitrogen inlet at the bottom and two interior Pyrex tubes which are cooled by circulating methanol at −20°C (Note 2). A Philips HPK 125-watt, high-pressure mercury lamp is placed inside the smaller Pyrex tube and the solution irradiated for 14 hours under a stream of dry nitrogen. The solvent is distilled off at normal pressure or by using a rotary evaporator and the residue dissolved in 100 ml of chloroform. After refluxing for 1 hour the formation of the azepine is complete, and the product is purified by column chromatography on silica gel (Note 3) with dichloromethane. The yellow fraction is collected, evaporated under vacuum, and the residual oil distilled at $5 \cdot 10^{-4}$ torr in a sublimation apparatus with a bath temperature of 50–60°C. The yield is 0.8 g (80%) of a yellow oil (Note 4).

2. Notes

1. The starting material is prepared in 45% yield by Diels-Alder addition of 1-acetylpyrrole with dimethylacetylenedicarboxylate.[1]
2. Cooling is necessary to prevent transformation of the azaquadricyclane to the azepine during illumination. If isolation of the photoproduct 3-acetyl-3-azadimethyltetracyclo[$3.2.0.0^{2,7}.0^{4,6}$] heptane-1,5-dicarboxylate is desired, the solution is concentrated under reduced pressure below 0°C

* Universität Freiburg, D-7800 Freiburg, Germany.
† Gorlaeus Laboratoria, Rijksuniversiteit, Leiden, The Netherlands.

to 80 ml and the tetracycle crystallized at −40°C after addition of 50 ml of cold methanol. The yield is 0.70 g of colorless crystals, dec. p. ~81°C.[2]

3. Fifty grams of silica gel (Woelm, 70–230 mesh) were used.

4. A short pathway and a high surface-to-volume ratio in the final distillation are recommended to prevent dimerization of the azepine which occurs on prolonged heating.[2]

3. Methods of Preparation

The procedure described is the only one known for the synthesis of 1-acetyl-4,5-dicarbomethoxy-1*H*-azepine.[2]

4. Merits of the Preparation

The photolysis of 7-azanorbornadienes, followed by thermolysis of the 3-azaquadricyclanes, is a rather general way to 4,5-disubstituted-1*H*-azepines according to the $3\sigma \rightarrow 3\pi$ route,[2] which is also successful in the analogous synthesis of 4,5-disubstituted oxepines.[3] 3,6-Disubstituted-1*H*-azepines are formed competitively if the ester groups are replaced by trifluoromethyl substituents.

References

1. R. Kitzing, R. Fuchs, M. Joyeux, and H. Prinzbach, *Helv. Chim. Acta*, **51**, 888 (1968).
2. H. Prinzbach, G. Kaupp, R. Fuchs, M. Joyeux, R. Kitzing, and J. Markert, *Chem. Ber.*, **106**, 3824 (1973).
3. W. Eberbach, M. Perroud-Arguëlles, H. Achenbach, E. Druckrey, and H. Prinzbach, *Helv. Chim. Acta*, **54**, 2579 (1971).

N-(1-ADAMANTYLMETHYLENE)-1-ADAMANTANAMINE

$$\text{1,3-di(1-adamantyl)aziridinone} \xrightarrow[-CO]{h\nu} \text{Ad–CH=N–Ad}$$

Submitted by ERACH R. TALATY,[1] AUBRY E. DUPUY, Jr., and LOUIS H. STEKOLL*
Checked by T. D. ROBERTS†

1. Procedure

A solution of 1,3-di(1-adamantyl)aziridinone[2] (0.20 g, 0.62 mmole) in 55 ml of pentane is placed in a cylindrical quartz tube (24 × 3.2 cm) fitted with a rubber septum, which is securely fastened with copper wire. Dry nitrogen is bubbled through the solution for 20 minutes and the latter is then irradiated (2537 Å) in an air-cooled Rayonet Srinivasan-Griffin chamber for 12 hours (Note 1). After the irradiation the solution has a cloudy appearance and is filtered. The clear filtrate is collected in a 100-ml, round-bottomed flask and the pentane is removed by means of a rotary evaporator. The residue is purified by sublimation (150°/0.2 mm) (Note 2) to give 0.14–0.15 g (76–82%) of the pure imine as white crystals, m.p. (sealed tube) 290–292° (dec.) (Note 3).

2. Notes

1. During the irradiation carbon monoxide is also produced. The evolution of this gas causes an increase in pressure and the septum expands. For this reason adequate precautions should be taken if the reaction is conducted on a larger scale. At the end of the irradiation the gas should be carefully vented in a hood.
2. The imine can be purified also by recrystallization from pentane.

* Wichita State University, Wichita, Kansas 67208.
† University of Arkansas, Fayetteville, Arkansas 72701.

3. A small amount of yellow solid (presumably a polymer) remains at the bottom of the sublimator.

3. Discussion

Because the aziridinone is prepared from adamantane-1-acetic acid[2] and the imine can be hydrolyzed to adamantane-1-carboxaldehyde, the foregoing preparation[3] constitutes a key step in a novel method for shortening the length of suitably substituted carboxylic acids by one carbon atom.

References

1. Part of this work was done at Louisiana State University in New Orleans, New Orleans, Louisiana 70122.
2. E. R. Talaty, A. E. Dupuy, Jr., and A. E. Cancienne, Jr., *J. Heterocycl. Chem.*, **4**, 657 (1967).
3. E. R. Talaty, A. E. Dupuy, Jr., and T. H. Golson, *Chem. Commun.*, **49** (1969).

2-AZA-*cis*-BICYCLO[2.2.0]HEX-5-EN-3-ONE

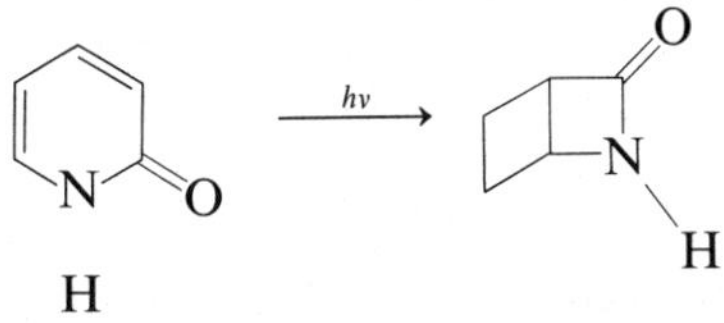

Submitted by WENDELL L. DILLING*
Checked by V. Y. MERRITT and R. SRINIVASAN†

1. Procedure

The ultraviolet radiation source is a 450-watt Hanovia medium-pressure mercury arc lamp, type 679A. The lamp is placed in a water-cooled 7740 Pyrex well (Note 1). The well is inserted in a cylindrical reaction vessel which has a 60/50 standard taper joint on top. The reaction vessel is constructed so that 2.1 liters of solution will cover the well to a level above the window of the lamp when in place (Note 2). The reaction vessel also contains a magnetic stirring bar and a rubber serum-capped side arm near the top for removing samples.

A solution of 2-pyridone (2-pyridinol; 2.00 g, 21.0 mmoles) (Note 3) in 2.1 liters of deionized water (10^{-2} *M*) is placed in the reaction vessel, stirred magnetically, and irradiated at ~25°. The course of the reaction may be followed by removing 1.0-ml samples periodically with a hypodermic syringe and diluting them with water to 5.0 ml. These solutions are analyzed for unreacted 2-pyridone in 1-mm cells by ultraviolet spectroscopy at 296 nm ($\varepsilon = 5400$) (Note 4). After 72 hours of irradiation the conversion of 2-pyridone is essentially complete.

The water is removed from the reaction solution under vacuum (10–20 mm) on a steam bath to give a solid which is sublimed at 60° (10 mm) to give 0.52–0.62 g (26–31%) of the white crystalline product, m.p. 68–69°; $\nu_{max}^{Fluorolube}$ 3220 (s), 3060 (w), 3045 (w), 3015 (m), 1710 (s), 1540 (w), 1340 (w) cm^{-1}; ν_{max}^{Nujol} 1285 (m), 1268 (s), 1198 (w), 1182 (w), 1138 (m), 1115 (s), 1048 (w), 980 (w), 969 (m), 938 (m), 901 (w), 865 (w), 828 (m), 774 (s), 729 (m), 702 (s), 676 (m), 496 (m) cm^{-1}; nmr ($CDCl_3$), a six-line multiplet centered at 4.15 (δ), a seven-line multiplet at 4.44, a four-line multiplet at 6.54, a nine-line multiplet at 6.64, and a broad multiplet at 6.2–7.3 ppm (NH) (one proton each) (Note 5). The sublimation residue contains the 4 + 4 dimer of 2-pyridone[2,3] (Notes 6 and 7).

* Environmental Sciences Research, The Dow Chemical Company, Midland, Michigan 48640.
† IBM Research Center, Yorktown Heights, New York 10598.

2. Notes

1. The lamp and quartz well with a tubular Pyrex filter, which may be used instead of the Pyrex well, are available from the Hanovia Lamp Division, Engelhard Hanovia, Inc., Newark, New Jersey.

2. The size of the reaction vessel may be varied in accordance with the amount of solution to be irradiated.

3. The 2-pyridone was commercial material (Aldrich Chemical Company) which had been purified by sublimation at 80–90° (0.2–0.4 mm), m.p. 107–108°.

4. The product does not absorb at 296 nm.

5. See ref. 1 for the nmr spectrum in $CDCl_3$ after D_2O exchange.

6. Use of a nitrogen purge with a 10^{-2} *M* solution of 2-pyridone at 20 ± 1° gave 1.18 g (59%) of purified product and ~30% of the dimer. The product can be prepared in higher yield (88%) by using a 10^{-3} *M* solution. However, the amount of material obtainable with the same volume of solution is smaller, 0.18 g, in 15 hours.

7. The product is a reasonably thermally stable material. At *ca.* 160° the neat melt undergoes a vigorous reaction to regenerate 2-pyridone.

3. Methods of Preparation

This photochemical reaction represents the only reported method for the preparation of the azabicyclohexenone. After this procedure was submitted the same reaction was published by other workers.[1]

4. Merits of the Preparation

The present process employs inexpensive, commercially available starting material, nonhazardous solvent, and a simple workup procedure. Related reactions of *N*-methyl-2-pyridone[4,5] and pyridine[6] itself have also been reported. The reaction can be adapted to some other substituted 2-pyridones.[1,7]

References

1. R. C. De Selms and W. R. Schleigh, *Tetrahedron Lett.*, 3563 (1972).
2. E. C. Taylor and R. O. Kan, *J. Am. Chem. Soc.*, **85**, 766 (1963).
3. L. A. Paquette and G. Slomp, *J. Am. Chem. Soc.*, **85**, 765 (1963).
4. E. J. Corey and J. Streith, *J. Am. Chem. Soc.*, **86**, 950 (1964).
5. L. J. Sharp IV and G. S. Hammond, *Mol. Photochem.*, **2**, 225 (1970).
6. K. E. Wilzbach and D. J. Rausch, *J. Am. Chem. Soc.*, **92**, 2178 (1970).
7. W. L. Dilling and N. B. Tefertiller, unpublished work.

BENZO[b]PHENANTHRO[9,10-d]FURAN

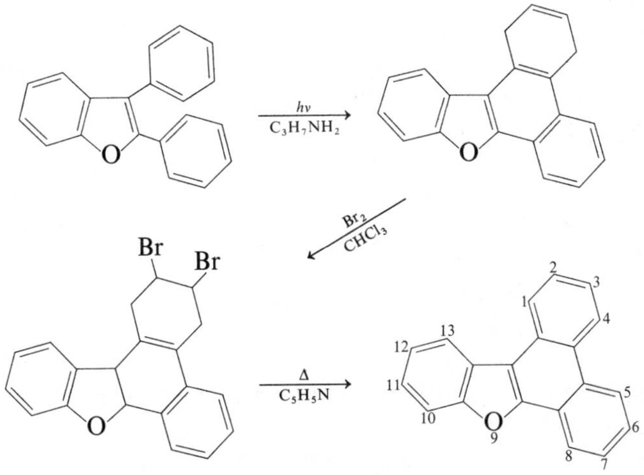

Submitted by A. COUTURE, A. LABLACHE-COMBIER, and H. OFENBERG*
Checked by J. CORNELISSE, G. M. GORTER-LAROY, and J. N. M. BATIST†

1. Procedure

A. 1,4-Dihydrobenzo(b)phenanthro[9,10-d]furan. A solution of 2,3-diphenylbenzo(b)furan (4.2 g, 0.015 mole) (Note 1) in 300 ml of propylamine is purged with nitrogen and irradiated in a quartz vessel for 12 hours with a Hanau NN 15-watt, low-pressure mercury lamp. The propylamine is removed by distillation in a vacuum and the residue washed with 30 ml of acetone. The solid reaction product is filtered and recrystallized twice from acetone. Crystalline needles of 1,4-dihydrobenzo(b)phenanthro[9,10-d]-furan (2.6 g, 65%) are obtained (m.p. 173°, sublimation at 132°).

B. 2,3-Dibromo-1,4-dihydrobenzo(b)phenanthro[9,10-d]furan. The 1,4-dihydrobenzo(b)phenanthro[9,10-d]furan (2.6 g, 0.01 mole) previously obtained is dissolved in 40 ml of chloroform and bromine (1.6 g, 0.5 ml) dissolved in 5 ml of chloroform is added with stirring. As soon as the color of bromine disappears crystalline needles of crude 2,3-dibromo-1,4-

* Laboratoire de chimie organique physique, Université des Sciences et Techniques de Lille, BP 36,59650 Villeneuve d'Ascq-France; H. Ofenberg is on leave from the University of Iasi-Rumania.

† Gorlaeus Laboratoria, Rijksuniversiteit, Leiden, The Netherlands.

dihydrobenzo(b)phenanthro[9,10-d]furan precipitate. After filtration they are washed with chloroform to yield 3.5 g (0.08 *M*) (Note 2).

C. Benzo(b)phenanthro[9,10-d]furan. Crude 2,3-dibromo-1,4-dihydrobenzo(b)phenanthro[9,10-d]furan (3.5 g, 0.08 mole) obtained previously is dissolved in 50 ml of pyridine. The solution is refluxed for 3 hours. During this period pyridine bromohydrate precipitates on the flask. After cooling the solution is diluted with 100 ml of water and stirred; 1 hour later the precipitate is filtered and washed with water. Benzo(b)phenanthro[9,10-d]-furan (1.9 g) is obtained. Its ethanolic solution is purified by charcoal. After filtration the pure compound is obtained by two crystallizations from ethanol. The yield is 1.7 g, m.p. 156° [40% with reference to 2,3-diphenylbenzo(b)furan, Note 3].

2. Notes

1. 2,3-Diphenylbenzofuran is easily obtained by condensation of benzoin with phenol by boric acid[1] or by cyclization in polyphosphoric acid of phenoxydesoxybenzoin.[2]
2. Pure 2,3-dibromo-1,4-dihydrobenzo(b)phenanthro[9,10-d]furan can be obtained by three recrystallizations of the crude material from benzene, m.p. 242°.
3. Benzo(b)phenanthro[9,10-d]furan can be obtained in one step from 2,3-diphenylbenzo(b)furan: 2 g of this compound dissolved in 300 ml of chloroform, when irradiated as in A for 8 hours, leads, after evaporation of the solvent, to a mixture of benzo(b)phenanthro[9,10-d]furan (52.5%) and the starting material. The reaction product can be separated from the starting material by preparative vpc (Autoprep A 700) on a 20 × 3/8 ft, 30% SE-30 column. Benzo(b)phenanthro[9,10-d]furan is also formed in low yield (12%) when the cyclization is performed in propylamine. It can be isolated from the mother liquor by filtration of the tars on a silica gel column after evaporation of the solvent and preparative vpc of the resulting mixture on the same column as above.

3. Merits of the Preparation

The method of preparation of benzo(b)phenanthro[9,10-d]furan is simple. It does not need chromatographic separation as does the direct photocyclization in chloroform. It can be carried out on a small (0.2 g) or large (5 g) amount of diphenylbenzo(b)furan. During this synthesis 1,4-dihydrobenzo(b)phenanthro[9,10-d]furan is obtained in good yield. The nonphotochemical preparation of phenanthro[9,10-d]benzo(b)furan, which has been described, is tedious.[3]

References

1. B. Arventiev and H. Ofenberg, *Studii si Cercet. Stiint. Chimie Iasi*, (1960), **11**, 305; C.A. **56**: 11554c.
2. Cl. Perrot and E. Cerutti, *C. Rend.Acad. Sci.*, (1967), **264**, 1301.
3. J. N. Chatterjea, V. N. Mehrotra, and S. K. Roy, *Ber.*, (1963), **96**, 1157.

BISADAMANTYLIDENE-1,2-DIOXETANE

$$\xrightarrow[\text{sensitizer}]{h\nu,\ O_2}$$

Submitted by W. ADAM*, J. -C. LIU*, J. STRATING†, J. H. WIERINGA†, and H. WYNBERG†
Checked by T.D. ROBERTS and T. WOOLDRIDGE‡

1. Procedure

The irradiation is carried out in a Pyrex test tube (20 mm O.D., 180 mm long); the top is capped with a serum stopper which supports a gas dispersion tube (Note 1) and a gas vent (Note 2). A solution of *bis*-adamantylidene (268 mg, 1.0 mmole) (Note 3) and tetraphenylporphorin (0.5 mg) (Note 4) in 20 ml of reagent grade carbon tetrachloride is placed in the reaction vessel, positioned vertically *ca.* 10 cm away from the light source. The latter consists of a DVY-650W Sylvania Tungsten-halogen lamp (Note 5), supported in a double-sleeved Pyrex tube (Note 6). Cooling water is passed through the inner sleeve while a filter of 0.2 *M* $CuCl_2$ and 0.5 *M* $CaCl_2$ filter solution (Notes 7,8) is circulated through the outer sleeve. With the oxygen gas purging through the solution at a slow rate (Note 9) and the light source activated, the singlet oxygen required is generated *in situ.*

Within 2 hours the photooxygenation is completed (Note 10). Removal of the solvent at reduced pressure and column chromatography (Note 11) of the crude product (300 mg, 100%) on silica gel (10 g) and eluting with hexane:chloroform (3:2) produce the pure 1,2-dioxetane (240 mg, 80%) as the first fraction, m.p. 173–4° (Note 11), colorless needles from methanol. A second crop (60 mg, 20%), admixed with sensitizer, can be obtained on further elution.

The proton nmr (CCl_4, TMS) shows complex multiplets at 1.30–2.25 (24*H*) and 2.40–2.75 (4*H*); the ir (CCl_4) shows a strong characteristic band at 920cm^{-1}. At reflux of an ethylene glycol solution adamantanone is formed quantitatively and accompanied by bright chemiluminescence (Note 12).

* University of Puerto Rico, Rio Piedras, Puerto Rico 00931.
† The University, Zernikelaan, Groningen, The Netherlands.
‡ University of Arkansas, Fayetteville, Arkansas 72701.

2. Notes

1. A tube with a medium porosity sintered glass disc is suitable.

2. A 22-gauge hypodermic needle was used.

3. This olefin (m.p. 184–5°) can be prepared by a variety of methods,[1,2] of which the one described in Reference 2 is the most convenient.

4. A generous gift sample was supplied by Dr. A. Adler. Other sensitizers and solvents can also be used.

5. This type of lamp is available in most photoshops that carry flash equipment. To permit maximum entry of the lamp into the cooling sleeve a quartz adapter (Model 514, Smith-Victor Corporation, Griffith, Indiana), which is also available in most photoshops, was used. The adapter was demounted from its base to permit extension of the electrical leads. The checkers used a Sylvania 500T30/Cl/U lamp that required a considerably longer irradiation time.

6. A single-sleeved tube can also be used, but then the filter solution must be circulated through a cooling bath to prevent excessive heat-up by the lamp.

7. A Masterflex Tubing Pump (Cole-Parmer Instrument Company) was found to be particularly useful for circulating the filter solution.

8. To serve as shield against undesirable infrared and ultraviolet radiation of the lamp.

9. Ordinary oxygen gas was used. A steady stream (20–30 ml/min) is essential to maintain a saturated solution because the dissolved oxygen is converted to singlet oxygen. A rapid oxygen flow rate is undesirable because it promotes solvent evaporation.

10. The progress of the reaction can be monitored by TLC on silica gel and hexane:chloroform (3:2) as the solvent system. Alternatively, the appearance of the 920cm^{-} absorption in the ir can be used.

11. Fast heating past its melting point leads to detonation. Although this material exhibits an unusually high thermal stability compared with its congeners[3] ($t^{1/2}$ about 30 minutes at 160°), as with other organic peroxides, it is cautioned to treat it as a potentially DANGEROUS chemical.

12. Quite generally four-membered ring cyclic peroxides chemiluminesce on thermal decomposition.[3,4]

3. Methods of Preparation

Bisadamantylidene-1,2-dioxetane has already been prepared[5] by cycloaddition of singlet oxygen, generated photochemically by methylene blue sensitization and chemically from the triphenylphosphite-ozone complex.

References

1. H. W. Geluk, *Synthesis*, 652 (1970).
2. A. P. Schaap and G. R. Faler, *J. Org. Chem.*, **38**, 3061 (1973).
3. K. R. Kopecky and C. Mumford, *Can. J. Chem.*, **47**, 709 (1969); T. Wilson and A. P. Schaap, *J. Am. Chem. Soc.*, **93**, 4126 (1971); N. J. Turro and P. Lechtken, *ibid.*, **95**, 264 (1973).
4. W. Adam and J. -C. Liu, *ibid.*, **94**, 2894 (1972); W. Adam and H. -C. Steinmetzer, *Angew. Chem. Int. Ed.*, **11**, 540 (1972); M. M. Rauhut, *Accounts Chem. Res.*, **2**, 80 (1969).
5. J. H. Wieringa, J. Strating, H. Wynberg, and W. Adam, *Tetrahedron Lett.*, 169 (1972).

bis(TRIFLUOROMETHYL)ACETOLACTONE

$$(CF_3)_2C(COF)_2 \xrightarrow[MeSO_3H]{H_2O_2} (CF_3)_2C(CO)_2O_2 \xrightarrow[-15°]{h\nu,\ CCl_4} (CF_3)_2C{-}C(=O){-}O$$

Submitted by W. ADAM, J. -C. LIU, and O. RODRIGUEZ*

1. Procedure

A. bis(Trifluoromethyl)malonyl peroxide. *bis*(Trifluoromethyl)malonyl difluoride (3.2 ml; 6.15 g; 25.2 mmoles) (Note 1) is condensed into a 50-ml round-bottomed flask with $\$$ 14/20 female joint, provided with a magnetic spin-bar, and kept below 10°C in an ice bath. While stirring magnetically and keeping the temperature below 10°C, add a solution of 98% H_2O_2 (4 ml) (Note 2) in methanesulfonic acid (10 ml) (Note 3) dropwise by remote control (Note 4). The reaction flask is protected from moisture by a silica gel drying tube and the contents stirred for 20 hours while maintaining the temperature below 10°C.

A vacuum pump is connected to the reaction flask by a $\$$ 14/20 adapter and, while stirring, the volatile materials are condensed into a Dry-Ice-cooled, demountable vacuum trap (Note 5). The product (lower layer of the condensate) is retrieved from the H_2O_2 (upper layer) with a disposable pipette and distilled from bulb to bulb (Note 6) to give 4.8 g of distillate. Distillation at atmospheric pressure affords 2.3 g (31%), b.p. 33–35°C at a bath temperature of 55°C, of unreacted malonyl difluoride as a first fraction and 1.2 g (20%), b.p. 62–63°C at a bath temperature of 85°C, of malonyl peroxide as a third fraction (Note 7).

The malonyl peroxide shows a characteristic carbonyl band at 1830 cm^{-1} (CCl_4) with a shoulder at 1870 cm^{-1} and a fluorine resonance at -62.9 ppm (CCl_4) from $CFCl_3$.

B. bis(Trifluoromethyl)acetolactone. The irradiation is carried out in a Pyrex test tube (15-mm O.D., 100 mm long) with $\$$ 14/20 female joint. The stoppered photo-vessel, containing a solution of the malonyl peroxide (119 mg, 0.5 mmole) in 2 ml spectrograde CCl_4, is placed into a transparent Dewar flask (Note 8), kept at -15°C by means of an ice-methanol mixture.

* University of Puerto Rico, Rio Piedras, Puerto Rico 00931.

The Dewar flask is positioned vertically *ca.* 10 cm away from the Pyrex-filtered beam of a Hanovia 500-watt mercury-xenon high pressure lamp (Note 9) to effect the photodecarboxylation.

Within 1 hour (Note 10) 81% (Note 11) of the malonyl peroxide is converted into the α-lactone, formed in 76% yield (Note 12) as a carbon tetrachloride solution. Attempts to isolate this material in the pure condensed phase leads to its destruction by decarboxylation and polymerization. The carbon tetrachloride solutions however, can be stored at Dry-Ice temperatures for weeks without noticeable decrease on the 1975/cm carbonyl band or δ −69.2 ppm (relative to $CFCl_3$) fluorine resonance.

2. Notes

1. A generous sample, b.p. 33° (760 mm) and m.p. 6°, was kindly supplied by Dr. D. C. England of Central Research, E. I. Dupont de Nemours Co., Wilmington, Delaware.

2. *Caution*! This hazardous chemical, available from FMC Corporation, Inorganic Chemicals Division, New York, should be handled with all safety measures (cf. FMC Bulletin No. 46).

3. Commercial material, available from Columbia Organic, b.p. 167° (10 mm) and m.p. 19–20°, was used.

4. A convenient technique for safe handling of concentrated H_2O_2 by remote control is to use the Cheng tube shown in the accompanying drawing,

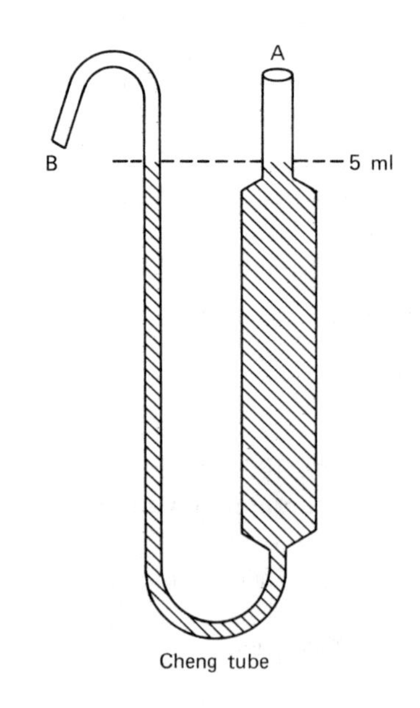

Cheng tube

developed by Dr. Y.-M. Cheng (M.Sc. Thesis, University of Puerto Rico, March 1968). The Cheng tube is charged with the H_2O_2 by using a disposable glass pipette through orifice A. Orifice A is then connected by a flexible rubber hose to a 10-ml plastic syringe, the piston of which is removed. After positioning the orifice B of the Cheng tube above the mouth of the reaction flask the piston is slowly introduced into the syringe barrel and the H_2O_2 transferred dropwise while standing behind a safety shield. In the same manner the H_2O_2-$MeSO_3H$ solution was first prepared and then transferred to the reaction flask.

5. Initially intermediate vacuum (30–40 mm) allows the reaction mixture to warm up slowly. At room temperature and 0.1-mm pressure for 45 minutes the residual volatile materials are removed. Alternative work-up by solvent extraction as in other malonyl peroxides[1] is undesirable in view of the great volatility of the product.

6. This bulb-to-bulb distillation was performed at room temperature and intermediate pressure (40–50 mm) was used to remove traces of H_2O_2.

7. It is not advisable to prolong the perhydrolysis time to increase the conversion because lower yields of less pure product are obtained. The recovered starting material can, however, be recycled in subsequent runs. The low recovery of malonyl difluoride is due to evaporation because long reaction times are required as a result of the resistance of the malonyl difluoride toward perhydrolysis compared with malonic acids.[1]

8. A rectangular quartz Dewar flask was used, but a cylindrical Pyrex Dewar flask will serve as well.

9. The lamp was housed in a SP-200 Bausch & Lomb Mercury Light Source, modified to accommodate the 500-watt Hanovia lamp.

10. In one instance a 1000-watt Schoeffel lamp was used and photodecarboxylation was complete in 16 minutes.

11. The progress of the decarboxylation can be visually followed until carbon dioxide gas evolution subsides. Spectroscopically the reaction can be monitored by ir, following the 1975 cm^{-1} α-lactone carbonyl and the 1830 cm^{-1} malonyl peroxide carbonyl bands. Photodecarboxylation to 100% conversion is not recommended because a lower yield of a α-lactone is obtained as a result of polymerization and decarbonylation.[2]

12. The yield was established by quantitative fmr, using *tetrakis* (trifluoromethyl)allene as an internal standard with δ -60.0 ppm relative to $CFCl_3$.

3. Methods of Preparation

bis(Trifluoromethyl)acetolactone has been prepared in a manner analogous to other α-lactones[1,2] by photodecarboxylation of *bis*(trifluoromethyl)-malonyl peroxide.[3]

References

1. W. Adam and R. Rucktäschel, *J. Am. Chem. Soc.*, **93**, 558 (1971); W. Adam and R. Rucktäschel, *J. Org. Chem.*, **37**, 4128 (1972).
2. O. L. Chapman, P. W. Wojtkowski, W. Adam, O. Rodriguez, and R. Rucktäschel, *J. Am. Chem. Soc.*, **94**, 1365 (1972).
3. W. Adam, J. -C. Liu and O. Rodriguez, *J. Org. Chem.*, **38**, 2269 (1973).

4,5-*bis*-TRIMETHYLSILOXYPENTACYCLO[4.3.0.0^{2,5}.0^{3,8}.0^{4,7}]-NONANE (4,5-*bis*-TRIMETHYLSILOXYHOMOCUBANE)

CO_2Me + Na + $(CH_3)_3SiCl$ $\xrightarrow[\Delta]{\text{toluene}}$ —$OSi(CH_3)_3$

CO_2Me $OSi(CH_3)_3$

(I) (II)

(II) $\xrightarrow[\text{cyclohexane}]{h\nu}$ $OSi(CH_3)_3$

$OSi(CH_3)_3$

(III)

Submitted by R. D. MILLER and D. L. DOLCE*
Checked by ERACH R. TALATY and LARRY M. PANKOW†

1. Procedure

A. 3,4-bis-trimethylsiloxytricyclo[*4.2.1.0*^{2,5}]*non-3,7-diene.* A dry 5-liter three-necked flask equipped with a Hershberg stirrer, condenser with drying tube, argon inlet, and 500-ml constant pressure addition funnel is charged with toluene (1300 ml) (Note 1) and clean sodium (52.8 g; 2.21 moles). The toluene is refluxed and the sodium dispersed by rapid stirring over a 1-hour period. After cooling the dispersion slowly to room temperature a solution containing *endo*-2,3-dicarbomethoxybicyclo[2.2.1]hept-5-ene[1] (113.9 g, 0.54 mole) and trimethylsilyl chloride (242 g, 2.23 moles) in 100 ml of toluene is added over 6 hours. The solution is then carefully heated (Note 2) to reflux for an additional 10 hours. After cooling, the reaction mixture is filtered through Celite and the toluene is removed by distillation at reduced pressure (40 mm). The crude material is distilled through a 6-in. Vigreux column and the main fraction (b.p. 78–81°, 0.05 mm, 80%) is used for further transformations. A small amount of high boiling material (b.p. 160–180°, 0.045 mm) is discarded. Glpc analysis (Note 3) indicates that the main fraction is >95% pure. The ir spectrum (neat) of (II) shows strong bands at 3060, 2960, 2920, 1710, 1300, 1250, 1210, 1170, 1085, 960, 900, 840, and 730/cm. The nmr spectrum of the product exhibits absorptions at

* IBM Research Laboratory, San Jose, California 95193.
† Wichita State University, Wichita, Kansas 67208.

$\tau(CDCl_3)$ 4.2 (unresolved t, 2*H*), 7.28–7.70 (m, 4*H*), 8.37 (ABq, $J = 8.5$Hz, 2*H*) and 9.90 (s, 18*H*).

B. 4,5-bis-Trimethylsiloxypentacyclo[$4.3.0.0^{2,5}.0^{3,8}.0^{4,7}$]*nonane*. A Vycor tube (33 × 3.8 cm) is charged with freshly distilled (II) (11.1 g; 0.038 mole) dissolved in 250 ml of dry spectrograde cyclohexane (0.15 *M*). The tube is sealed with a syringe stopper and degassed with a stream of dry nitrogen for 15 minutes. The solution is placed in a photochemical reactor (Note 4) and irradiated for 48 hours at 30°. The progress of the reaction is conveniently monitored by infrared spectroscopy (Note 5). After this time the starting material is completely consumed (Note 6), as determined by infrared analysis and the nmr spectrum of a concentrated aliquot. The solvent is removed on a rotary evaporator (100 mm, 40°) and the mobile yellow residue distilled (b.p. 85–88°, 0.12 mm) through a 6-in. vacuum jacketed Vigreux column to yield 9.24 g (84%) of colorless product. This material is shown by glpc analysis (Note 7) to be >95% pure. The ir of (III) shows strong bands at 2980, 1355, 1305, 1255, 930, 905, 875, and 840 cm^{-1}. The 100 MHz nmr spectrum reflects the high symmetry and shows the following absorptions: $\tau(CCl_4)$ 6.85 (s, 6*H*, 1/2wd 2.3Hz), 8.22 (s, 2*H*, 1/2wd 2.7Hz) and 9.75 (s, 18*H*). The six-proton singlet apparently results from an accidental chemical equivalence of the bridgehead methine protons (Note 8).

2. Notes

1. The toluene was carefully dried immediately before use by distillation from sodium. All other liquid reagents were purified by distillation.
2. Care must be taken in reheating the mixture after the addition because a strong but short-lived exothermic reaction takes place suddenly. After it subsides the reactants may be refluxed without difficulty overnight. A color change from gray to deep bluish-purple accompanies the exothermic process.
3. Analyses were carried out on a Hewlett Packard Model 5750 analytical gas chromatograph using a 6 ft × 1/8 in. column packed with 10% UCW98 on Chromosorb W (HMDCS washed) operated at 150°.
4. The photochemical reactor, model RPR-208, equipped with eight U-shaped low-pressure mercury lamps whose estimated output at 2537 Å is 120 watts, is available from the Southern New England Ultraviolet Company. The sample tube itself was air cooled by a powerful fan mounted in the base of the reactor. The checkers performed the reaction on a fivefold smaller scale in an RPR-100 reactor made by the same company.
5. The reaction progress is conveniently monitored by a decrease in the starting material absorption at 1710 cm^{-1}. The photolysis solution may be

used without dilution by using 0.1 mm sodium chloride cavity cells and a cyclohexane blank. The photochemical cycloaddition has been run successfully on a smaller scale in anhydrous ether, but the solvent volatility makes infrared monitoring more difficult.

6. Overirradiation does not appear to be seriously detrimental; however, the photolysis should be run until the starting material is completely consumed because it is extremely difficult to separate it from the product by conventional techniques. For this reason a final check of a concentrated aliquot of the irradiation mixture by nmr is recommended before workup.

7. The analyses performed are described in Note 3.

8. The use of dry pyridine as the nmr solvent causes considerable broadening and the development of some fine structure at the base of the low-field, six-proton absorption at 6.85τ.

3. Merits of the Preparation

This preparation of 4,5-*bis*-trimethylsiloxyhomocubane, which is based on a published procedure,[2] offers a direct, high-yield route to previously unknown 4,5-disubstituted homocubyl derivatives. The strong absorption of the *bis*-trimethylsiloxycyclobutene chromophore in the starting material permits the cycloaddition to be performed without the usual photosensitizers. The thermal stability of the trimethylsiloxy linkage facilitated the purification of the product while, at the same time, it is reactive toward various nucleophiles and nucleophilic solvents. This also allows the removal of the trimethylsilyl groups under extremely mild conditions and the preparation of a number of other interesting derivatives.

References

1. M. S. Morgan, R. S. Tyson, A. Lowy, and W. E. Baldwin, *J. Am. Chem. Soc.*, **66**, 404 (1944).
2. R. D. Miller and D. L. Dolce, *Tetrahedron Lett.*, 4541 (1972).

CARBOMETHOXYCYCLOÖCTATETRAENE
(Methyl Cycloöcta-1,3,5,7-Tetraene-1-carboxylate)

$$\text{C}_6\text{H}_6 + HC{\equiv}CCOOCH_3 \xrightarrow{h\nu} \text{C}_8\text{H}_7COOCH_3$$

Submitted by GARY L. GRUNEWALD and JOSEPH M. GRINDEL*
Checked by JOSE A. ORS†

1. Procedure

The apparatus consists of a cylindrical irradiation vessel made of quartz, 38 cm long and 7.6 cm O.D., to one end of which a ground glass joint is attached (Note 1). The source of ultraviolet (2537 Å) radiation is a Rayonet Reactor (Note 2).

The vessel is charged with a solution of methyl propiolate (75 ml, 0.88 mole) (Note 3) in 1500 ml of reagent-grade thiophene-free benzene and is connected to a Friedrich condenser equipped with a drying tube. The vessel is centered between the lamps of the reactor and the contents are irradiated for 20 hours with magnetic stirring (Note 4). The yellow solution is decanted into a 2-liter, round-bottomed flask and the unreacted benzene and methyl propiolate are codistilled at 79–80°, leaving a reddish oil as a residue. Two methods for the purification of this residue are described below. The clear distillate is returned to a clean quartz vessel (Note 5) and irradiated for an additional 20 hours. This procedure is repeated for 15–25 cycles (Note 6). After the last cycle has been completed (Note 7) the combined residues are stripped of any remaining low-boiling material *in vacuo* to produce 103–140 g.

The highest yield of the desired ester is obtained by chromatography of the residue in 30-g portions on a 300-g silica gel column (4.0 × 59.5 cm) slurry packed with 5% ether/hexane (Note 8) and eluted with the same solvent. The ester, which moves down the column as a yellow band, is the first product eluted. In a typical 25-cycle run 140 g of residue, which yielded 59 g (37%) of ester after chromatography, was obtained.

A simpler and more rapid purification scheme, but one that produces a lower yield of purified ester, is to add the residue to 10 volumes of hexane.

* Department of Medicinal Chemistry, School of Pharmacy, University of Kansas, Lawrence, Kansas 66045.

† IBM Watson Research Center, Yorktown Heights, New York, 10598.

An insoluble reddish oil immediately separates and the supernatant solution of the ester is decanted through a funnel containing Celite 545 (Note 9); it is then dried over calcium chloride and the solvent is removed *in vacuo* to yield 58.8 g of crude ester from a typical 20-cycle run. After distillation (65–73°/1.5 mm Hg) a clear yellow oil (29.9 g, 21%) (Note 10) is obtained: b.p. 50.0–50.5° (0.3 mm) [lit.[2] 89° (0.4 mm)], n_D^{28} 1.5934 (lit.[2] 1.5938); ir (neat) 3020 (HC═C), 1710 (C═O), 1635 (C═C), 695 (C═C), and 650 (C═C) cm^{-1}; nmr ($CDCl_3$) δ 6.9–5.7 (m, 7, ring C*H*) and 3.75 (s, 3, OCH_3); mass spectrum (70 eV) *m/e* (rel. intensity) 162 (28), 131 (12), 103 (100), 102 (47), 78 (12), 77 (59), and 51 (32).

2. Notes

1. Vessel RQV-323 is supplied by the Southern New England Ultraviolet Co., Middletown, Connecticut 06457. Two such containers are necessary (see Note 5).
2. Model RPR-208 (supplied by the Southern New England Ultraviolet Co.) fitted with RUL-2537 lamps. The quartz vessel is placed in the center over a magnetic stirrer.
3. Supplied by Farchan Research Laboratories, 4702 East 355th Street, Willoughby, Ohio 44094.
4. Product yield decreases beyond this time.
5. The reaction vessel is coated with a yellow polymer at the end of a cycle. The polymer is removed by treatment with 200–300 ml of chromic acid cleaning solution (5 g $Na_2Cr_2O_7$/5 ml water/100 ml H_2SO_4 conc.), flushing with water, washing with a warm detergent solution, rinsing with distilled water, and drying.
6. Recycling is arbitrarily discontinued at this point because the yield of residue from a cycle at the end of this time is only about 50% that of the first cycle.
7. It is not necessary to proceed through so many cycles if a smaller amount of the ester is desired.
8. Skellysolve B (b.p. 60–68°) is used whenever "hexane" is mentioned.
9. Supplied by Fisher Scientific Company.
10. Percentage yields are based on the amount of methyl propiolate used.

3. Methods of Preparation

Carbomethoxycyclooctatetraene has been previously synthesized by esterification of cyclooctatetraene carboxylic acid[2] and by irradiation of ·nzene and methyl propiolate.[1,3] A modification of the latter procedure is ·zed herein.

4. Merits of the Synthesis

This route offers the advantage of not requiring cyclooctatetraene, a relatively expensive compound, as a starting material. The original photocyclization synthesis of Bryce-Smith[1,3] was a single-cycle process with very low yield which required an elaborate quartz wool scrubber to keep the surface of the immersion well clean during the photolysis. Although it may seem that the current procedure is time consuming because of the number of cycles required, it actually takes only an hour or so each day of personnel time.

Carbomethoxycyclooctatetraene is a useful intermediate for the preparation of a variety of substituted cyclooctatetraenes. We have prepared cyclooctatetraenylmethyl alcohol in nearly quantitative yield by lithium aluminum hydride reduction. This alcohol can be converted in high yield with MnO_2 to cyclooctatetraenylcarboxaldehyde. The latter compound is a versatile synthetic intermediate; we have, for example, prepared the cyanohydrin of this aldehyde.

References

1. D. Bryce-Smith and J. E. Lodge, *J. Chem. Soc.*, 695 (1963).
2. A. C. Cope, M. Burg, and S. W. Fenton, *J. Am. Chem. Soc.*, **74**, 173 (1952).
3. D. Bryce-Smith and J. E. Lodge, *Proc. Chem. Soc.*, 333 (1961).

1-CHLORO-1-METHYLCYCLOPROPANE

$$CH_2{=}C(CH_3){-}CH_2Cl \xrightarrow{h\nu} \text{1-chloro-1-methylcyclopropane } (C_3H_4(CH_3)Cl)$$

Submitted by STANLEY J. CRISTOL and RANDALL J. DAUGHENBAUGH*
Checked by T. D. ROBERTS and C. T. ROBERTS†

1. Procedure

Acetonitrile (600 ml, spectrophotometric grade), acetone (160 ml, spectrophotometric grade), and β-methylallyl chloride (6.0 ml) (Note 1) are placed in an 800-ml immersion well equipped with a magnetic stirrer, nitrogen bubbler, and reflux condenser. The quartz well is water-cooled and has a 450-watt Hanovia lamp provided with a Corex filter (Note 2). Nitrogen is bubbled through the solution continuously during the preparation. After one-half hour of purging irradiation is begun. Three 6.0-ml portions of β-methylallyl chloride are added, each after a 10-hour interval (Notes 3 and 4). Irradiation is continued until, at 50 hours, all the starting material is consumed. The reaction can be followed by gc (Note 5) or nmr (Note 6).

The reaction is worked up by dissolving 20 ml of decane in the reaction mixture, followed by repeated washings with water to remove the acetone and acetonitrile solvents (Note 7). The resulting 35 ml of solution is dried over anhydrous potassium carbonate and distilled. 1-Chloro-1-methylcyclopropane is collected from 42–48° at 625 torr. The yield is 6.6 g (30%) of 1-chloro-1-methylcyclopropane of at least 94% purity (Note 8).

2. Notes

1. Common impurities are water, which is removed by drying over anhydrous potassium carbonate and distilling, and a small amount of 1-chloro-2-methylpropane which is inert in the photoreaction.
2. A Pyrex filter permits the reaction to occur but at a slower rate.
3. The major side reaction is dimerization (Note 6) and the best yield is realized when the concentration of β-methylallyl chloride is kept low. The checkers ran this reaction on half the scale described here.

* Department of Chemistry, University of Colorado, Boulder, Colorado 80302. Research supported by Public Health Service Grant CA-13199 from the National Cancer Institute.
† University of Arkansas, Fayetteville, Arkansas 72701.

4. The initial rate of disappearance of methylallyl chloride under the described conditions is approximately 1 ml/hr. The reaction slows as irradiation proceeds, presumably because of quencher buildup, so that each portion takes longer to react.

5. Starting material and product can be analyzed on a Carbowax gc column. On a 1/8 in $\times$ 4-meter 25% Carbowax-20M column at 95°, flow ~30 ml/min, the following retention times in minutes are observed: 1-chloro-1-methylcyclopropane, 3.0; acetone, 3.5; β-methylallyl chloride, 5.0; acetonitrile, 10.

6. With nmr, starting material, product, and dimeric by-product are all easily measured. A 2-ml aliquot is removed from the photolysis mixture and 0.5 ml of carbon tetrachloride (1% TMS) is added and stirred. The solution is then washed with water and placed in an nmr tube. The useful analytical peaks for each compound follow: β-methylallyl chloride—$2H$ at 3.97δ, 1-chloro-1-methylcyclopropane—$3H$ at 1.63δ, 5-chloro-4-chloromethyl-2,4-dimethyl-1-pentene—$4H$ at 3.49δ.

7. A small amount of NaCl in the wash water will aid in the washing process. It is important that washing be complete, for this is the easiest point to remove the solvents. The checkers found that seven washings were required.

8. Impurities are starting material and the 1-chloro-2-methylpropane present in starting material.

3. Methods of Preparation

This preparation is based on a general reaction described earlier.[1] Cyclopropanes are generally made by addition of carbenes to double bonds or by γ-elimination reactions; halocyclopropanes may be made by direct chlorination or by the Hunsdiecker reaction or one of its modifications.

4. Merits of Preparation

This is a general method[1] for the photoisomerization of β-alkylallyl chlorides to 1-chloro-1-alkylcyclopropanes, for that of α- and γ-alkylallyl chlorides to 2-chloro-1-alkylcyclopropanes, and for α- and γ-arylallyl chlorides to 2-chloro-1-arylcyclopropanes. It is not at all successful with β-arylallyl chlorides. 1-Chloro-1-alkylcyclopropanes are not readily available by other reactions.

References

1. (a) S. J. Cristol and G. A. Lee, *J. Am. Chem. Soc.*, **91**, 7554 (1969); (b) S. J. Cristol, G. A. Lee, and A. L. Noreen, *ibid.*, **95**, 7067 (1973).

(20R)-AND-(20S)-20-CHLORO-5α-PREGNANE-3β,16β-DIOL

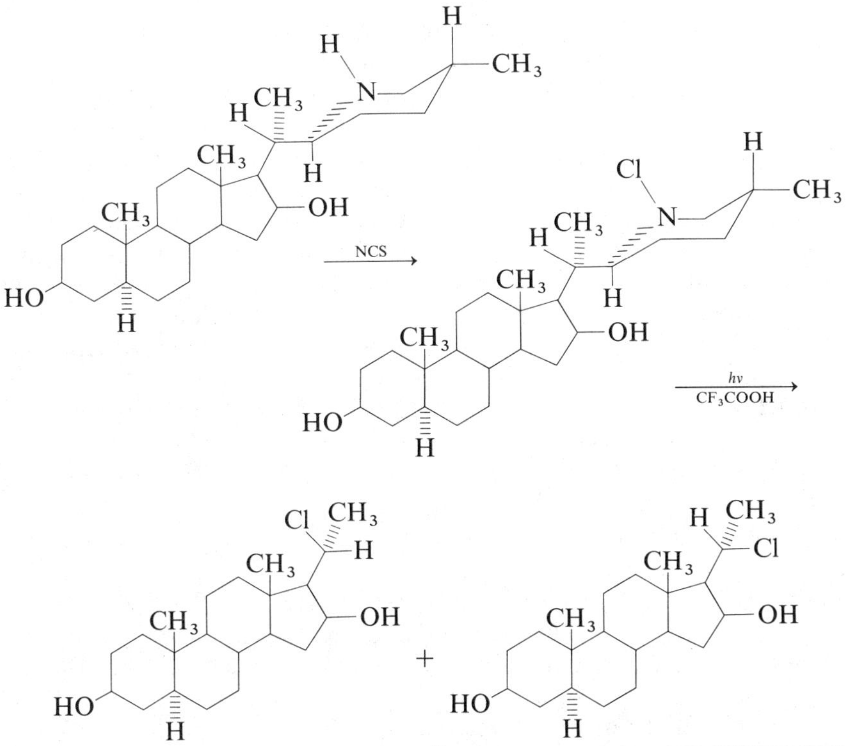

NCS

$h\nu$ / CF_3COOH

+

Submitted by GÜNTER ADAM and KLAUS SCHREIBER*
Checked by J. CORNELISSE and G. M. GORTER-LAROY†

1. Procedure

(22S,25R)-22,26-Epimino-5α-cholestane-3β,16β-diol (Tetrahydrosolasodine, 3.65 g, 0.0087 mole, Note 1) is suspended in 750 ml of methylene chloride (Note 2) and cooled to −5 to −10° in an ice-salt bath. At this temperature a solution of 1.17 g (0.0087 mole) *N*-chlorosuccinimide (Note 3) in 100 ml of methylene chloride is added during 30 minutes under stirring. After stirring for an additional 30 minutes without the cooling bath the clear solution is washed three times with water (1-l) and dried over anhydrous

* Institute for Plant Biochemistry of Research Centre for Molecular Biology and Medicine, Academy of Sciences of the DDR, Halle/S., DDR.
† Gorlaeus Laboratoria, Rijksuniversiteit, Leiden, The Netherlands.

sodium sulfate; the drying agent is then removed by filtration. Evaporation of the methylene chloride solution at room temperature and reduced pressure (rotary evaporator) in a 250-ml quartz flask gives 3.6–3.8 g of a white-crystalline *N*-chloro derivative of m.p. 275° (dec.) which is used directly for the photolysis (Note 4). The crude *N*-chloroamine is dissolved in 40 ml of ice-cold trifluoroacetic acid (Note 5), and the quartz flask is equipped with a gas inlet tube and a reflux condenser. A slow stream of argon cooled in an ice-salt bath is bubbled through the solution to push away air and oxygen and to keep the photolysis temperature below 30°. After 10 minutes of deoxygenation time the photolysis is begun (Note 6). It is usually complete after 15–20 minutes of irradiation (Note 7). Removal of the trifluoroacetic acid at reduced pressure and room temperature produces a yellow-brownish oil which is dissolved in 40 ml of methylene chloride. Anhydrous $NaHCO_3$ is added under stirring until the solution is no longer acidic (Note 8). The salt is removed by filtration and washed three times with additional methylene chloride (100 ml). Evaporation of the combined organic phases under reduced pressure gives 3.6–4.0 g of amorphous product, which is dissolved in 15 ml of benzene and chromatographed on 270 g of alumina (Note 9). Elution with benzene-ether 8:2 v/v yields first 0.6–0.8 g (Note 10) of the less polar (R_F 0.48, brownish-violet, Note 11) crystalline (20S)-20-chloro-5α-pregnane-3β,16β-diol, m.p. 218–222° (dec.) showing after recrystallization from methanol m.p. 225–226° (dec.) and $[\alpha]_D^{21}$ + 39.8° (c = 0.463, pyridine). Further elution with benzene-ether 8:2 gives 0.5–0.8 g mixture of (20S)- and (20R)-stereoisomers (R_F 0.48 and 0.42) followed by 0.6–0.9 g (Note 10) of pure crystalline, more polar (R_F 0.42, violet) (20R)-20-chloro-5α-pregnane-3β,16β-diol of m.p. 200–203° (dec.) which shows after recrystallization from methanol m.p. 214–215° (dec.) and $[\alpha]_D^{19}$ + 0.8° (c = 0.441, pyridine) (Notes 12–14).

2. Notes

1. This starting steroidal amine is readily available *via* catalytic hydrogenation of the spirosolane alkaloid solasodine[1] or sodium borohydride reduction of the corresponding 5α-alkaloid soladulcidine.[2]

2. The methylene chloride was purified by shaking with anhydrous $NaHCO_3$, filtration, and distillation before use.

3. Freshly crystallized from chloroform-hexane before use, m.p. 149° (dec.).

4. The melting point is not sharp and depends on the rate of heating. After recrystallization from chloroform-hexane the *N*-chloro compound shows m.p. 280° (dec.) and $[\alpha]_D^{22}$ − 50.0° (c = 0.396, chloroform).[3]

5. The trifluoroacetic acid is freshly distilled before use.

6. The uv-source was a U-shaped high-pressure mercury lamp (500 watt) type Th U 500 (VEB THELTA-Elektroapparate, Zella-Mehlis, DDR), arranged externally below the quartz flask at a distance of 20 cm. The quartz flask was cooled from the outside with a fan.

7. The progress of the photofragmentation is monitored by removing one drop of the solution being irradiated and adding one drop of a 5% potassium iodide solution and one drop of starch solution. Absence of the blue color indicates a corresponding absence of chlorine and the end of the radical chain reaction.

8. This $NaHCO_3$ treatment is for neutralization of the remaining traces of fluoroacetic acid. The end point of this process could be detected by direct test of the solution with Unitest indicator paper.

9. Neutral alumina of grade III (Brockmann scala) was used with the following typical elution sequence (50-ml fractions): Fractions 1–10 with benzene, 11–18 with benzene-ether 9:1, 19–47 with benzene-ether 8:2 (column size 3.4 × 27 cm).

10. The yields are somewhat variable and moved between 20–25% for the pure (20S)- and 23–30% for the (20R)-compounds, based on starting (22S,25R)-22,26-epimino-5α-cholestane-3β,16β-diol. They could be further increased by rechromatography of the mixture fractions containing both unseparated epimers.

11. The course of the column chromatography is followed by TLC on silica gel G (Merck), using chloroform-methanol 9:1 v/v for development and vanillin-phosphoric acid (0.1 g vanillin in 10 ml of 50% phosphoric acid, after spraying for 10 minutes at 100°) for detection.

12. For further physical data of both 20-chloropregnanes see ref. 4.

13. Similar photofragmentations leading to modified 20-chloropregnanes of this type have been also described by us.[4,5]

14. The identification of the 5-methyl-Δ^1-piperidine fragment originating in this photofragmentation has been reported.[6]

References

1. H. Rochelmeyer, *Arch. Pharm.*, **277**, 329 (1939).
2. G. Adam and K. Schreiber, *Z. Chem.*, **9**, 227 (1969).
3. K. Schreiber and G. Adam, *Liebigs Ann.*, **666**, 155 (1963).
4. G. Adam and K. Schreiber, *Tetrahedron*, **22**, 3581 (1966).
5. G. Adam and K. Schreiber, *Liebigs Ann.*, 2048 (1973).
6. G. Adam, *Chem. Ber.*, **101**, 1 (1968).

endo-7-CYANO-2,3-BENZOBICYCLO[4.2.0]OCTA-2,4-DIENE

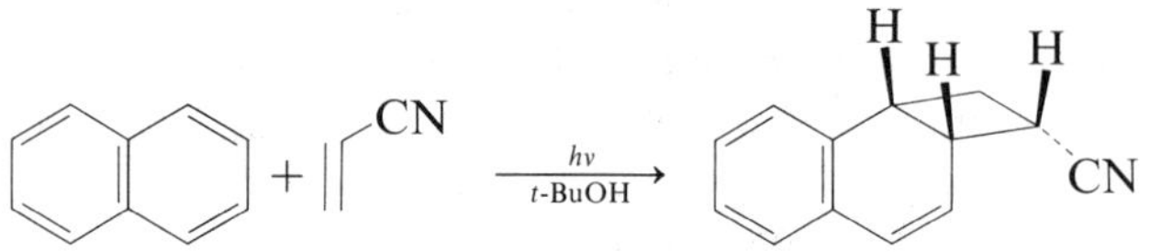

Submitted by R. M. BOWMAN, C. W. HUANG and J. J. McCULLOUGH*
Checked by J. CORNELISSE and G. M. GORTER-LAROY†

1. Procedure

A solution of naphthalene (1.025 g, 0.008 mole) and acrylonitrile (21.2 g, 0.4 mole) (Note 1) in 400 ml of 1:1 *tert*-butyl alcohol-*iso*-propyl alcohol mixture is prepared. The solution is irradiated in a pyrex photoreactor, fitted with an inlet tube for admission of argon, a port for sampling, and a centrally located, water-cooled quartz immersion well which contains the lamp and pyrex sleeve (Note 2). The solution is purged with argon for 20 minutes and irradiated for 4 hours by which time GLC monitoring shows that the product concentration is not increasing significantly (Note 3). The solutions from three such (identical) runs are combined and evaporated (Note 4). The crude mixture is extracted with 5:1 hexane-ether mixture (250 ml) and filtered through celite; the filtrate is then evaporated. The polymer-free mixture is chromatographed on 55 × 2.0 cm of silica gel (Note 5), slurry-packed in and eluted with benzene; 200-ml fractions are collected. Fractions 16–29 contain *endo*-7-cyanobenzobicyclo[4.2.0]octa-2,4-diene, 1.8 g with m.p. 97–98° (lit.[1] 97–98°) on crystallization from light petroleum (Note 6).

2. Notes

1. Acrylonitrile was Eastman Practical Grade, purified according to Bevington and Eaves[2], and redistilled before use, b.p. 77.5°.
2. A Hanovia 450-watt mercury vapor lamp was used. See *Organic Photochemical Syntheses*, Vol. 1, 1971, p. 14, for an illustration of a similar apparatus.
3. GLC was on 5 ft × 1/8 in. of QF-1 on Chromosorb W at 170°; the retention time of the product was 3.4 minutes.[1]
4. In our experience repeating the run is preferable to increasing the scale in each run.

* McMaster University, Hamilton, Ontario, Canada.
† Gorlaeus Laboratoria, Rijksuniversiteit, Leiden, The Netherlands.

5. Silica gel was Grace, grade 923 (100–200 mesh).

6. Minor products and naphthalene are found in chromatography fractions other than 16–29.

References

1. R. M. Bowman, C. Calvo, J. J. McCullough, R. C. Miller, and I. Singh, *Can. J. Chem.*, **51**, 1060 (1973).
2. J. C. Bevington and D. E. Eaves, *Trans. Faraday Soc.*, **55**, 1777 (1959).

endo-7-CYANO-6-METHOXY-2,3-BENZOBICYCLO[4.2.0]-OCTA-2,4-DIENE

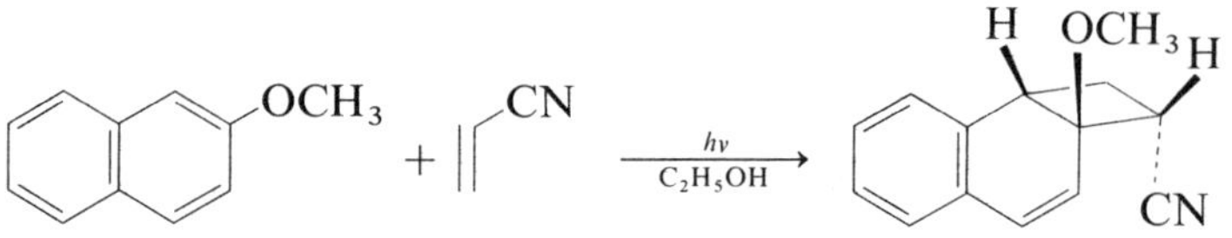

Submitted by T. R. CHAMBERLAIN, R. C. MILLER, and J. J. McCULLOUGH*
Checked by T. D. ROBERTS†

1. Procedure

A solution of 2-methoxynaphthalene (6.0 g, 0.0379 mole) (Note 1) and acrylonitrile (30.0 g, 0.565 mole) (Note 2) in 1 liter of absolute ethanol is prepared. The solution is irradiated in a pyrex photoreactor and fitted with an inlet tube for admission of argon, a port for sampling, and a centrally located, water-cooled quartz immersion well which contains the lamp and pyrex sleeve (Note 3). The solution is purged with argon for 30 minutes and irradiated for 12 hours (Note 4). The solution is evaporated and extracted with ether (4 × 100 ml). The extracts are filtered through celite and evaporated; the crude mixture is chromatographed on a 60 × 4.0 cm silica gel column, slurry packed in, and eluted with benzene-hexane (4:1) (Note 5). Fractions (250-ml) are collected, and fractions 13–20 contain *exo*- and *endo*-7-cyano-6-methoxy-2,3-benzobicyclo[4.2.0]octa-2,4-diene, 3.67 g.

The ratio of the latter is approximately 1:2, by GLC analysis (Note 4). Fractional crystallization (three times) of this mixture from ether-light petroleum (Note 6) gives 2.50 g of *endo*-7-cyano-6-methoxy-2,3-benzocyclo-[4.2.0]octa-2,4-diene with m.p. 85–86° (lit.[1] 85–86°).

2. Notes

1. 2-Methoxynaphthalene obtained from Eastman Organic Chemicals had m.p. 72–72.5° from 95% ethanol.

2. Acrylonitrile was Eastman Practical Grade, purified according to Bevington and Eaves[2] and redistilled before use, b.p. 77.5°.

3. A Hanovia 450-watt mercury vapor lamp was used. See *Organic Photochemical Syntheses*, Vol. 1, 1971, p. 14, for an illustration of a similar apparatus.

* McMaster University, Hamilton, Ontario, Canada.
† University of Arkansas, Fayetteville, Arkansas 72701.

4. GLC analysis on 5 ft × 1/8 in. SE-30 on Chromosorb W at 190° shows 2-methoxynaphthalene at 2 minutes, the *endo*-cycloadduct at 6 minutes and the *exo*-cycloadduct at 6.5 minutes. The checker observed that a white solid was precipitated during irradiation. This was filtered and discarded.

5. Silica gel was Grace, grade 923 (100–200 mesh).

6. The *exo*- and *endo*-adducts interconvert thermally and heating during crystallization is to be avoided.

References

1. T. R. Chamberlain and J. J. McCullough, *Can. J. Chem.*, **51**, 2578 (1973).

2. J. C. Bevington and D. E. Eaves, *Trans. Faraday Soc.*, **55**, 1777 (1959).

cis-1,2-CYCLOBUTANE DICARBOXYLIC ACID ANHYDRIDE

$$C_2H_4 + \text{maleic anhydride} \xrightarrow[C_6H_5COCH_3-CH_3CO_2C_2H_5]{h\nu} \text{cis-1,2-cyclobutanedicarboxylic anhydride}$$

Submitted by JORDAN J. BLOOMFIELD and DENNIS C. OWSLEY*
Checked by R. SRINIVASAN†

1. Procedure

The apparatus consists of a 12-liter, round-bottomed flask with a flanged central neck (Note 1), fitted with a water-cooled quartz immersion well (Note 2) which is tightly seated in a 20-mm-thick Teflon disk tapered to fit the flask clamp. The disk is drilled to pass 7-mm tubes for gas inlet, gas exit (to hood), and sampling (Note 3). A Pyrex filter (Note 4) containing a 1000-watt GE lamp (Note 5) is placed in the well.

The flask is filled with a mixture of maleic anhydride (490 g, 5.0 moles), ethyl acetate (10.6 liters), and acetophenone (35 g) (Notes 6, 7), and the assembled apparatus is clamped securely. The apparatus, wrapped in aluminum foil (Note 8), is immersed in a refrigerated cooling bath at -5° or below. Ethylene is bubbled through the gas dispersion tube (Note 9) at 1–1.5-liters/min (Note 10). After 1.5–2 hours the flow is reduced to 0.2 liter/min and the lamp is turned on (Note 11, 12). The course of the cycloaddition is followed by GLC (Note 13) at 24-hour intervals. After 6 1/2–7 days (Note 14) maleic anhydride is reduced to a very low or zero level and the light is turned off. The reaction mixture is concentrated to about 800 ml under reduced pressure (about 100 mm) on a steam bath (see Note 6). The *hot* residue is filtered on a coarse-sintered disk funnel to remove a white precipitate, 27–53 g (Note 15), which is washed with hot ethyl acetate. The combined filtrates are distilled at 4–8 mm through a 25-cm vacuum jacketed Vigreux column fitted with a variable take-off head (Note 16). The early fractions, 35–60 g, distilling from 70–105°, contain acetophenone, maleic anhydride (depending on the completeness of the photolysis), and small

* Corporate Research Department, Monsanto Company, 800 North Lindbergh Boulevard, St. Louis, Missouri 63166
† IBM Research Center, Yorktown Heights, New York 10598

amounts of the product anhydride. A small intermediate fraction (about 10 g, b.p. 105–115°) contains acetophenone and product. Cyclobutane dicarboxylic acid anhydride is collected at 126–129°/5 mm in two fractions which solidify in the receiver. The combined weight is 437–446 g (70–78%). The purity determined by GLC is more than 99% (see Note 13), although the last 50–75 g is yellowish in color. Further purification is effected by crystallization from 1:1 benzene-cyclohexane. A black, brittle, glassy residue, 80–90 g, remains in the pot.

2. Notes

1. For example, a Kontes reaction flask, Catalog TG-40, No. 612250, Kontes Glass Company, Vineland, New Jersey.
2. The dimensions of the well are inner wall, 70 mm; outer wall, 84 mm; inner depth, 440 mm; outer length, 460 mm. The water inlet tube, 5 mm, reaches nearly to the bottom with the external tubulatures of 8-mm quartz.
3. Nylon bushings, Ace Glass catalog 600 No. 5029 (Ace Glass Co., Vineland, New Jersey). The holes are drilled completely through the disk with a 5/16-in. bit. Then each hole is enlarged with a 27/64-in. bit to slightly more than halfway through the disk and tapped with a 1/2–13 standard screw tap.
4. A test tube (not a sleeve) 65 mm I.D. and 390–490 mm long.
5. A GE-H1000A36-15 lamp with the outer envelope removed is used in conjunction with a GE lamp ballast LDAS, Model 9T64Y166. The checker used a Rayonet type RS lamp in an external irradiation mode.
6. Maleic anhydride was Fisher certified material. Ethyl acetate was Mallinckrodt reagent grade which was recovered and used without further treatment in subsequent photolyses. Acetophenone was Matheson, Coleman and Bell material. The checker used one hundredth of this scale in a reaction vessel that was proportionately smaller.
7. Benzophenone, 40 g, was used as a photosensitizer with equally good results. Acetophenone is preferred because it is easier to remove in the volatile fraction preceding the product. The distilled product always contains codistilled benzophenone when it is used and recrystallization becomes a necessity.
8. The foil wrap is necessary to protect the eyes of people working in the vicinity.
9. The dispersion tube is a coarse or extra coarse cylindrical tube centered under the immersion well. A flat, coarse disk is satisfactory.

10. Gas flow is controlled with a Bantam Meter, Catalog TG-40, No. K627950, Style 10, Model 1, tube size 6, Kontes Glass Co., Vineland, New Jersey.

11. It is important that an adequate supply of water be maintained to the immersion well throughout the photolysis.

12. A slow flow of nitrogen is maintained inside the lamp well. The nitrogen protects the metal parts of the lamp and prevents ozone formation. It must not, however, be vigorous enough to cool the lamp or lamp efficiency will be severely reduced.

13. Any of the variety of columns can be used. A column 3.1 m long and 3.2 mm I.D. with either 1% OV-17 or 1% OV-225 on 100/120 mesh Chromosorb G programmed from 70° at 10°/min gives excellent separations.

14. After 48–72 hours the solution is slightly cloudy. This does not appear to impede the photolysis. No residue buildup is found on the immersion well. Sometimes the dispersion tube becomes clogged which prevents an adequate ethylene flow. The tube can be unclogged by increasing the gas flow to 1-liter/min for several minutes. It is helpful to do this at least once a day.

15. This appears to be mainly a polymer (m.p. >330°) but may contain some maleic anhydride photodimer.

16. The product has a tendency to solidify in the head and delivery tube. A heat lamp aimed at the head and the top of the collection flask will prevent this solidification.

3. Methods of Preparation

1,2-Cyclobutane dicarboxylic acid anhydride has previously been prepared by a multistep synthesis via adipic acid[1] or by a thermal, low-conversion dimerization of acrylonitrile,[2] followed by hydrolysis and dehydration of the diacid to the anhydride.[3] The photolysis process may also be conducted at −65°.[4] With a low concentration of maleic anhydride, about 0.2 *M*, the yield is 90%.[5]

4. Merits of Preparation

The reaction is easily adaptable to a wide variety of maleic anhydride derivatives and olefin partners but is particularly valuable with ethylene as a partner.[4] Although the time of irradiation is fairly long, minimum attention is required during the irradiation. Product isolation is simple and the yield is very good. The submitters have used the same apparatus, with reduced irradiation times, for simple batch preparation of 500–800-g quantities of the 1-chloro and 1,2-dichloro 1,2-cyclobutane dicarboxylic acid anhydrides.

References

1. E. R. Buchman, A. D. Reims, T. Skei, and M. J. Schlatter, *J. Am. Chem. Soc.*, **64**, 2696 (1942).
2. E. C. Coyner and W. S. Hillman, *ibid*, **71**, 324 (1949).
3. E. Vogel, O. Roos, and K. -H. Disch, *Ann.*, **653**, 55 (1962).
4. D. C. Owsley and J. J. Bloomfield, *J. Org. Chem.*, **36**, 3768 (1971).
5. J. J. Bloomfield and D. C. Owsley, unpublished work.

cis-3,4-CYCLOBUTENEDICARBOXYLIC ACID ANHYDRIDE

$$\text{maleic anhydride} + C_2H_2 \xrightarrow[CH_3CO_2C_2H_5,\ -65^\circ]{h\nu,\ C_6H_5COCH_3} \text{cis-3,4-cyclobutenedicarboxylic acid anhydride}$$

Submitted by JORDAN J. BLOOMFIELD and DENNIS C. OWSLEY*
Checked by R. SRINIVASAN†

1. Procedure

The apparatus consists of a 4-liter reaction kettle (Note 1) fitted with a coarse (or extra coarse) gas dispersion tube, cooling coil, gas exit tube, low-temperature thermometer, thermocouple lead (Note 2), and a three-wall Pyrex immersion well (Note 3) containing a 1000-watt GE lamp (Note 4). The fittings are introduced through holes drilled in a 27-mm-thick Teflon disk 175 mm in diameter (Note 5).

The assembled apparatus is filled *via* a sampling hole with a mixture of maleic anhydride (49 g, 0.5 mole), ethyl acetate (2800 ml) and acetophenone (14 g) (Note 6). Nitrogen at 50 ml/min is bubbled through the dispersion tube. The entire kettle is cooled in a large container of solid carbon dioxide/acetone and the water is turned on in the inner well (Note 7). When the temperature reaches -60°C, the acetylene is passed in at 8 liters/min for 15–20 minutes (Note 8). The flow is then reduced to about 0.2 liter/min of acetylene and 0.05 liter/min of nitrogen and the lamp is turned on (Notes 9, 10). The solution is maintained at $-65 \pm 10^\circ$ (Note 11) and is sampled at intervals for GLC analysis (Note 12). After 9.5 hours the lamp is turned off and the solvent is stripped at 100 mm. The residue is distilled at 5 mm through a 25-cm vacuum jacketed Vigreux column fitted with a variable take-off head. The first fraction, 14 g, b.p. 65–85°, contains acetophenone, maleic anhydride, and a trace of cyclobutene dicarboxylic acid anhydride. The second fraction, b.p. 85–106°, 0.2–0.7 g, is mainly product anhydride contaminated with acetophenone. The third fraction, 43–45 g (69–72%), b.p. 106–112°, is essentially pure anhydride (Notes 13, 14, 15).

* Corporate Research Department, Monsanto Company, 800 North Lindberg Boulevard, St. Louis, Missouri 63166.

† IBM Research Center, Yorktown Heights, New York 10598.

2. Notes

1. The basic apparatus has been described.[1] The checker conducted the reaction on 1/20th of this scale in a cylindrical vessel which had a concentric inner well that was filled with coolant.

2. The thermocouple is connected to a thermocontroller which operates a relay connected to the house nitrogen supply. The house nitrogen then forces liquid nitrogen from a storage Dewar through the cooling coil. If a low-temperature controller is not available, one can be instantly constructed by simply reversing the thermocouple leads. The instrument must then be calibrated or the set point adjusted so that the desired temperature is maintained.

3. The well is constructed as follows: three Pyrex test tubes are constructed of 81 mm O.D., 60 mm I.D., and 47 mm I.D. tubing, 370 mm, 400 mm, and 390 mm long, respectively. The two larger tubes are joined by a ring seal. A capillary tube is attached about 15 mm below the ring seal for later evacuation of this portion of the apparatus. A bead, 5 mm wide, is fashioned in the outer wall. This prevents the well from falling through the Teflon support ring (Note 5) and permits ready removal of the well for cleaning if that becomes necessary. Near the top at opposite sides of the inner tube are attached two 8-mm O.D. tubulatures. To one of these, on the inside of the tube and very close to the wall, is connected a tee-tube joined to two vertical tubes of 4-mm I.D. which reach almost to the bottom. This is the water inlet. Now the third and smallest diameter tube is attached. The outermost container is then evacuated to at least 5×10^{-6} mm (the higher the vacuum attainable the better), baked at 150°, and finally sealed. (Similar containers have been constructed of quartz. See detail drawing.)

4. The lamp is a GE-H1000A36-15 with the outer envelope removed. It is used in conjunction with a GE lamp ballast LDAS-Model 9T64Y166. In order for it to fit into the well the top metal disk is trimmed with sheet metal shears and the side support wires are bent in closer to the lamp. The checker used a Rayonet type RS lamp and irradiated the sample externally.

5. In the center of the disk a hole 82.0 mm in diameter is drilled. Another hole 90.1 mm in diameter and 15 mm deep is then cut. Finally, a third hole 91.9 mm in diameter and 10 mm deep is cut. An O-ring, size 2-238 (Vitone or Neoprene), is placed around the well up to the bead and the well is inserted in the Teflon disk. The O-ring is seated firmly by pushing it into place with the blunt blade of a thin spatula or even a pocket knife. The cooling coil (an 8-mm Pyrex coil with 2-cm spacing and 7-mm tubes

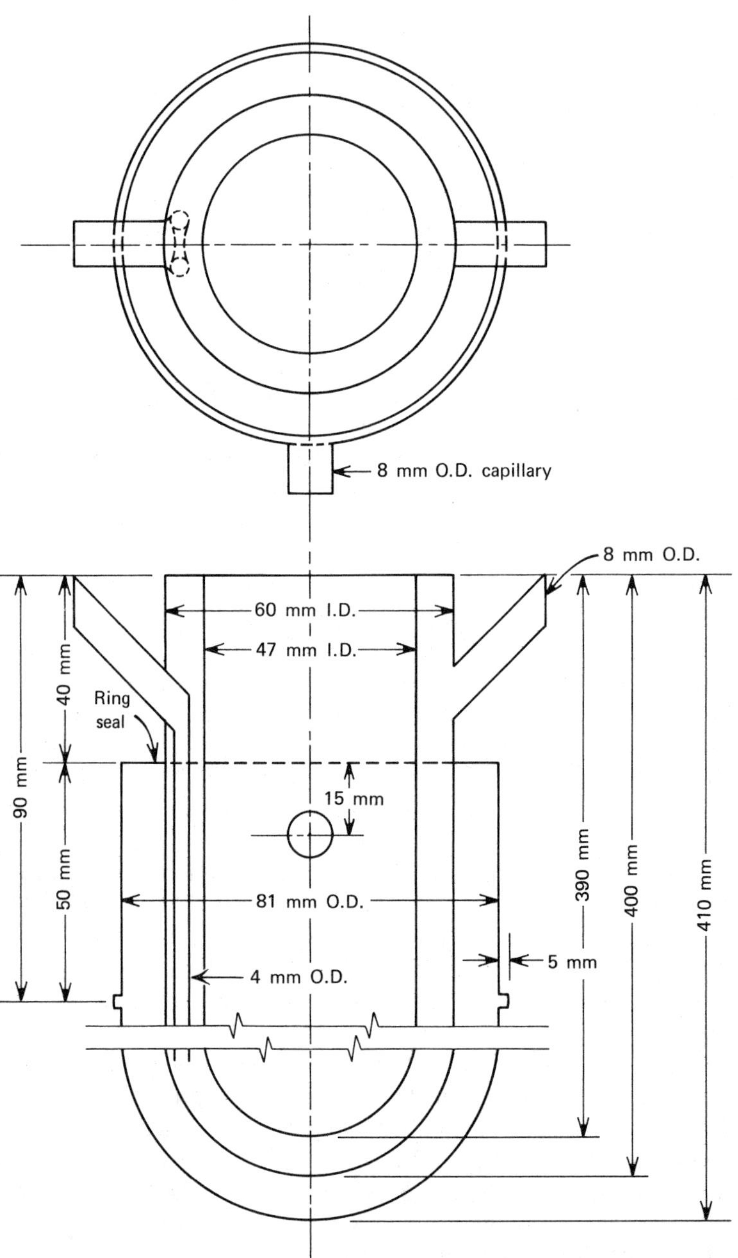
Triple-walled Dewar
8 mm O.D. capillary
8 mm O.D.
60 mm I.D.
47 mm I.D.
40 mm
Ring
seal
90 mm
15 mm
50 mm
81 mm O.D.
390 mm
400 mm
410 mm
5 mm
4 mm O.D.

sealed at both ends or a stainless steel coil 6.35-mm O.D., which provides better heat transfer and is not subject to breakage), gas dispersion tube, and thermometer are introduced through 5/16-in. holes drilled entirely through the disk. A 27/64 inch hole is then drilled on the same center 18 mm deep. This hole is threaded with a 1/2–13 standard screw tap. Each tube extending through the disk is of 7-mm Pyrex or has a 7-mm section sealed on the end. Tight seals are made with Nylon bushings, Ace Glass Catalog 600 No. 5029 (Ace Glass Co., Vineland, New Jersey).

6. The maleic anhydride was Fisher certified. Ethyl acetate was Mallinckrodt reagent grade. Acetophenone was obtained from Matheson, Coleman and Bell. Acetylene was ordinary welding grade used with no purification.

7. The water flow should be maintained at a brisk rate even when the lamp is turned off. This is especially true with quartz apparatus. The unsilvered vacuum jacket is not sufficient to prevent standing water from freezing (particularly in the quartz wells, which appear to go "soft" quite rapidly).

8. It is important to monitor the level of the liquid in the reaction kettle carefully. At this low temperature acetylene is essentially miscible and it is possible to overflow the reaction vessel.

9. It is advantageous to keep a slow flow of nitrogen in the lamp compartment and to surround the reaction kettle with aluminum foil to protect the eyes of people working in the laboratory.

10. The gas flow is controlled by a set of ganged Kontes Bantam meters, Catalog TG-40 No. K627950 of the appropriate flow rate ranges, Kontes Glass Co., Vineland, New Jersey.

11. It may be convenient to add solid carbon dioxide to the cooling bucket from time to time to conserve liquid nitrogen.

12. A column 3.1-m long and 3.2-mm I.D. packed with either 1% OV-17 or 1% OV-225 on 100/120 mesh Chromosorb G programmed from 70° at 10°/minute gives excellent separations.

13. The anhydride tends to crystallize in the head and delivery tube. A heat lamp aimed at the head will eliminate this difficulty.

14. The pot residue is mainly polymer. It is possible to isolate a small amount of a mixture of isomers of bicyclopropyl tetracarboxylic anhydride as the tetramethyl ester by refluxing the pot residue with methanol and a catalytic amount of concentrated sulfuric acid.

15. The size run described appears to be optimal. When larger scale runs are tried or when larger amounts of maleic anhydride are used but added in incremental amounts, larger quantities of product are produced, but the percent yield is lowered and the rate begins to slow down.

3. Methods of Preparation

The photochemical addition of acetylene to maleic anhydride is the most efficient and direct route to this compound.[2] The photocyclization of muconic anhydride has also been described.[3]

4. Merits of Preparation

The reaction conditions described permit the preparation of an interesting compound in very good yield. In addition, the apparatus can be used for a wide variety of low-temperature (or even elevated-temperature) photolytic reactions; for example, the same general conditions applied to the addition of ethylene to maleic anhydride gives yields of 90%.

The 1000-watt GE street lamp shortens reaction times by a factor of 6 to 10 compared with the 450-watt lamp in general use for preparative work.

References

1. D. C. Owsley and J. J. Bloomfield, *Org. Prep. Proc. Int.*, **3**, 61 (1970).
2. G. Koltzenburg, P. G. Fuss, and J. Leitich, *Tetrahedron Lett.*, 3409 (1966); W. Hartmann, *Chem. Ber.*, **102**, 3974 (1969), **104**, 2864 (1971).
3. G. J. Fonken, *Chem. Ind.*, 1575 (1961).

CYCLOMERISATION OF 7,7′-PENTAMETHYLENE-DIOXYCOUMARIN

$h\nu$, $\lambda > 335$ nm

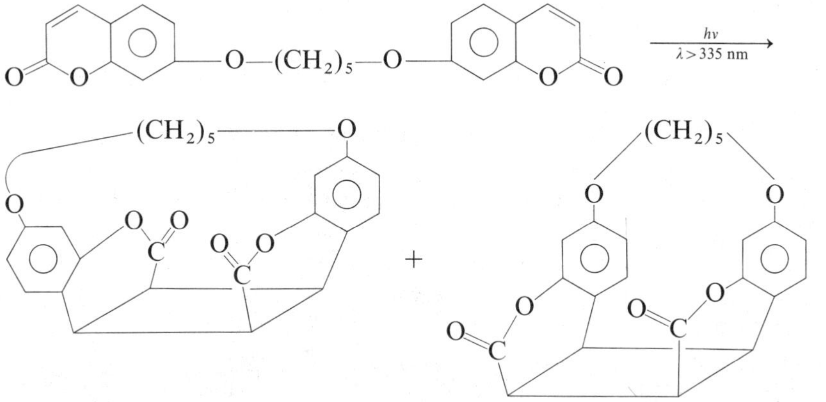

Submitted by H. LOOS, L. LEENDERS and F. C. DE SCHRYVER*
Checked by J. CORNELISSE and J. N. M. BATIST†

1. Procedure

A. To a solution of 7-hydroxycoumarin (9.72 g, 0.06 mole) in 750 ml of dry acetone (Note 1) in the presence of anhydrous potassium carbonate (16.6 g, 0.22 mole), stirred magnetically in a flask protected from moisture, add a solution of 1,5-dibromoalkane (8.5 g, 0.03 mole) in 50 ml of the same solvent[1] (Note 2). The mixture is refluxed for 12 hours. The acetone is distilled off and the residue mixture is transferred into an excess of water. The precipitate is filtered and washed with distilled water. The crude product is dissolved in acetone and precipitated with diethyl ether. This is repeated twice to remove the monofunctional bromo compound.

The dioxycoumarin is further purified by column chromatography on Florisil 60–100 mesh eluting with chloroform. The solvent is removed on a rotary evaporator and the residue recrystallized several times in benzene. The purity of the product is checked by TLC (silicagel, chloroform with 5% acetone as developing solvent). Yield 72%. The pure product melts at 138–8.5°C and is identified by ir, nmr, and mass spectroscopy.

B. The irradiation of a solution of 7,7′-pentamethylenedioxycoumarin (0.50 g) in CH_2Cl_2 (132 ml) is carried out in a pyrex vessel under nitrogen or argon atmosphere in a Rayonet photochemical reactor, type RPR 208, fitted

* K. Universiteit Leuven, Celestijnenlaan 200 F, B-3030 Heverlee, Belgium.
† Gorlaeus Laboratoria, Rijksuniversiteit, Leiden, The Netherlands.

with a set of eight RUL 3500 Å lamps. All the light below 335 nm is cut off by use of a liquid filter[2] (Note 3). The irradiation is continued until no further decrease in the starting material can be observed by thin layer chromatography, using Florisil (Merck) and chloroform-acetone (90:10) ~ 20 hours.

The cyclomers have an *endo* structure. A nmr spectrum of the crude mixture after evaporation of the solvent allows the calculation of the relative percentages of the two cyclomers, based on the absorption of the phenyl protons: 42% head-to-head and 58% head-to-tail cyclomer.

On fractional crystallization from dry benzene 0.14 g (34.6%) head-to-head isomer is separated. The head-to-head isomer is recrystallized from benzene (Note 4).

The filtrate is evaporated to dryness. A nmr spectrum of this residue shows the presence of the head-to-tail isomer, some oligomeric species (5%), and a small amount of the head-to-head isomer (7%). The head-to-tail isomer is separated by chromatography on a column filled with Florisil (60–100 mesh) eluting with chloroform. The head-to-tail isomer is recrystallized from toluene (Note 4). The column is freed of organic material by eluting with glacial acetic acid; the fraction contains a small amount of head-to-head isomer and the oligomeric species. The acid is evaporated and the residual product is extensively washed with water. The purity of the isomers is checked by TLC (silica gel; chloroform with 5% acetone as developing solvent). The structure of the isomers is proved by ir, nmr, and mass spectroscopy. The dipole moment of the head-to-head isomer equals 6.7 D, whereas the head-to-tail isomer has a dipole moment of 2.0 D.[3]

2. Notes

1. The acetone was dried over potassium carbonate for about four days.
2. Other 7,7′-polymethylenedioxycoumarins, with methylene units between two and nine, can be prepared in a manner analogous to 7,7′-pentamethylenedioxycoumarin. Irradiation of these dioxycoumarins lead to the same isomers, but the isomer ratio is a function of the chain length.
3. The liquid filter contains a solution of 479 g of sodium bromide and 3 g of lead acetate or 3 g of lead nitrate in 1 liter of water. To cut off the light below 335 nm a filter solution with a path length of 1 cm is used.
4. The head-to-head isomer sublimes at 236.5–238.5°C and the head-to-tail isomer at 291.5–294°C.

References

1. L. H. Leenders, E. Schouteden, and F. C. De Schryver, *J. Org. Chem.*, **38**, 957 (1973).
2. M. P. Rappoldt. Thesis, University of Leiden, 1958.
3. P. Huyskens and F. Cracco, *Bull. Soc. Chim. Belges*, **69**, 422 (1960).

5,7-DI-*tert*-BUTYL-3,3-DIMETHYL-(3*H*)INDOLE-1-OXIDE

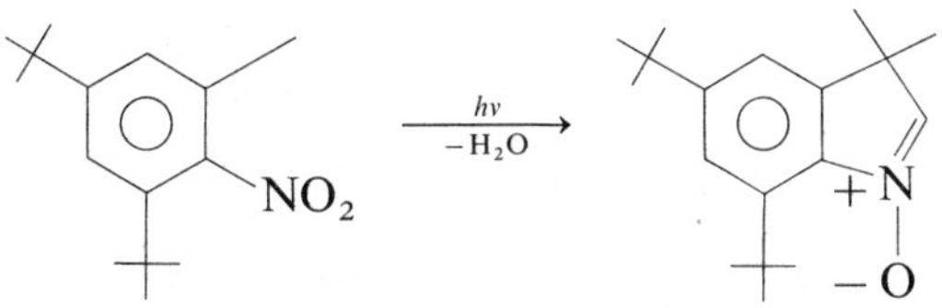

Submitted by DIETRICH DÖPP*
Checked by J. CORNELISSE and G. M. GORTER-LAROY†

1. Procedure

1,3,5-Tri-*tert*-butyl-2-nitrobenzene (50 g lot, 0.17 mole) (Note 1) is finely powdered and spread on a 50 × 30 cm reflecting rectangular tin pan and irradiated with two sunlight lamps (Note 2) mounted above; the distance between the crystal layer and the periphery of the bulbs is 40 cm. The crystals soon assume a yellow-brown color at their exposed surfaces. After 45 minutes the photolysis is discontinued, the crystals are collected, thoroughly ground, and spread again for another 45-minute period of irradiation. This operation is repeated once more.

After a total of 135 minutes of irradiation the slightly sticky crystals (48.2 g, Note 3) are dissolved in 500 ml of warm cyclohexane. This solution is transferred to a 5-cm-wide column slurry-packed with a 20-cm-long layer of deactivated silica gel (Note 4) in cyclohexane. The same solvent is used for elution of the starting material, and a 1700-ml fraction containing 42.8 g (86% recovery, Note 5) is collected. Minor by-products are subsequently eluted with 500 ml of benzene, followed by 250 ml of benzene/ethyl acetate 5:1.

As soon as a small sample of the eluate gives a deep blue color on adding an equal volume of methanol and one drop of a 1% aqueous solution of ferric chloride, a 300-ml fraction (benzene-ethyl acetate 5:1, fraction A), containing 2.0 g of a mixture of 5,7-di-*tert*-butyl-1-hydroxy-3,3-dimethyl-(3*H*)-2-indolene (major component) and the title compound, is collected. The bulk of the title compound is eluted with pure ethyl acetate (1500 ml, fraction B), and a crop of 2.5 g of material is thus obtained.

Fractions A and B, separately, may be further purified by preparative TLC with 5:1 benzene/ethyl acetate (Note 6). Pure 5,7-di-*tert*-butyl-3,3-dimethyl-(3*H*)indole-1-oxide has m.p. 166° (dec.).

* Fachbereich Chemie, Universitat Kaiserslautern, 675 Kaiserslautern, Germany.
† Gorlaeus Laboratoria, Rijksuniversiteit, Leiden, The Netherlands.

2. Notes

1. See ref. 1 for preparation. Material of m.p. 202–204° was used by the submitter.

2. Two new Ultra-Vitalux 300-watt lamps supplied by Osram GmbH were used. Both ir and near uv radiation, as emitted by this kind of lamp, seem to be necessary for successful irradiation. The checkers used an identical light source.

3. The decrease in weight of the material is due mainly to mechanical losses and the sublimation of the starting material.

4. Silica gel (0.2–0.5-mm particle diameter) is slurried with water, suction filtered, and air dried.

5. Because the desired product is more light-sensitive than the starting material, higher conversions are accompanied by substantial secondary photolysis.

6. The usual technique[2] of preparative TLC is followed. Two 40 × 20 cm plates, covered with a 2-mm layer of silica gel, were used for both fractions. Plates of the type 60 F 253 available from Merck or SIL G 200 UV 254 available from Macherey & Nagel were found to be equally suitable. The zones were detected by quenching the fluorescence of the indicator added to the adsorbent and eluted with acetone.

3. Methods of Preparation

The method described here is a modification of a published[3] procedure.

References

1. P. D. Bartlett, M. Roha, and R. M. Stiles, *J. Am. Chem. Soc.*, **76**, 2349 (1948).
2. H. Halpaap, *Chem.-Ing.-Tech.*, **35**, 488 (1963).
3. D. Döpp and K. -H. Sailer, *Chem. Ber.*, **108**, 301 (1975).

1,2-DICARBOMETHOXYCYCLOÖCTATETRAENE

$$C_6H_6 + CH_3OOC\text{-}C{\equiv}C\text{-}COOCH_3 \xrightarrow{h\nu} C_8H_6(COOCH_3)_2$$

Submitted by LEO A. PAQUETTE and RONALD S. BECKLEY*
Checked by T. D. ROBERTS and C. T. ROBERTS†

1. Procedure

A cylindrical quartz photolysis vessel (Note 1) is charged with benzene (500 ml) and dimethyl acetylenedicarboxylate (15 ml, 17.5 g, 0.123 mole). This solution is irradiated with a bank of 16 2537 Å lamps in a Rayonet reactor (Note 2) for six 8-hour periods (Note 3). The benzene is removed on a rotary evaporator to give a yellow residue which, on distillation, returns 11.4 g (65%) of unreacted dimethyl acetylenedicarboxylate, b.p. 84–86° (6 mm, Note 4). The residual reddish oil or waxy solid is purified by chromatography on neutral alumina; elution with benzene-hexane (1:1) gives 4.0–4.4 g (16.4–18.2%; Note 5) of 1,2-dicarbomethoxycycloöctatetraene. Recrystallization of this material from approximately 10 ml of methanol produces large yellow prisms, m.p. 107–109°.

2. Notes

1. The cylinder measures 6 cm. in diameter and is 25 cm. long. At the top it carries a 24/40 joint for mounting a condenser if desired.
2. Model RPR-100 supplied by the Southern New England Ultraviolet Co., Middletown, Connecticut.
3. The operating temperature of the solution is approximately 50°. After the 8-hour period the polymer, which develops on the inner surface of the vessel, must be removed to permit incursion of light. Alcoholic potassium hydroxide solutions have been found to be effective cleansers. The checkers did not find this polymeric residue during irradiation.
4. This material is pure enough to be recycled. The last traces of the acetylenic diester can be removed by lowering the pressure to approximately 1.5 mm.

* The Ohio State University, Columbus, Ohio 43210.
† University of Arkansas, Fayetteville, Arkansas 72701.

5. On the basis of recovered dimethyl acetylenedicarboxylate, a conversion of about 50% has taken place at this point.

3. Methods of Preparation

The procedure described is a modification of that described in the literature.[1,2] 1,2-Dicarbomethoxycycloöctatetraene has also been prepared by a copolymerization process.[3]

4. Merits of the Preparation

Ultraviolet irradiation of solutions of appropriately substituted acetylenes in benzene provides a general synthetic entry to otherwise difficultly accessible cyclooctatetraene derivatives.

References

1. E. Grovenstein, Jr. and D. V. Rao, *Tetrahedron Lett.*, 148 (1961); E. Grovenstein, Jr., T. C. Campbell, and T. Shibata, *J. Org. Chem.*, **34**, 2418 (1969).
2. D. Bryce-Smith and J. E. Lodge, *J. Chem. Soc.*, 695 (1963).
3. J. E. Meili, Ph.D. Thesis, Massachusetts Institute of Technology, 1952; A. C. Cope and H. C. Campbell, *J. Am. Chem. Soc.*, **73**, 3536 (1951), **74**, 179 (1952); A. C. Cope and D. S. Smith, *ibid.*, **74**, 5136 (1952).

1,2-DIHYDRONAPHTHO[2.1-d]-3*H*-INDANO[2,1-f]-*N*-PHENYLSUCCINIMIDO[4,3-b]-7α-THIA-2β,3β,5α,6β-TETRAHYDROBICYCLO[2.2.1]HEPTA-2,5-DIENE

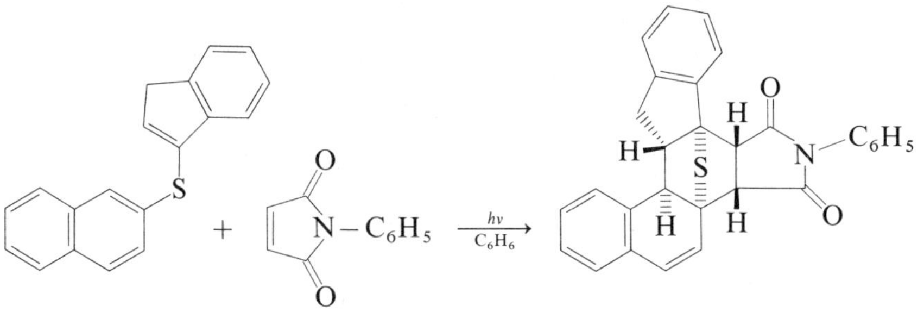

Submitted by A. G. SCHULTZ and M. B. DeTAR*
Checked by L. CATE*

1. Procedure

A. 3-(β-Naphthyl)indenyl sulfide. A solution of β-naphthalenethiol (11.2 g, 0.0700 mole) (Note 1), 1-indanone (9.25 g, 0.0700 mole) (Note 1), and *p*-toluenesulfonic acid (~50 mg) in benzene (50 ml) is treated as described.[1] Two crystallizations of the crude product from chloroform-methanol give 3-(β-naphthyl)indenyl sulfide (15.9 g, 83%), m.p. 98–99°.

B. 1,2-Dihydronaphtho[2,1-d]-3H-indano[2,1-f]-N-phenylsuccinimido[4,3-b]-7α-thia-2β,3β,5α,6β-tetrahydrobicyclo[2.2.1]hepta-2,5-diene (Note 2). A solution of 3-(β-naphthyl)-indenyl sulfide (5.00 g, 0.0182 mole, 0.0520 *M*), *N*-phenyl maleimide (Note 3, 7.90 g, 0.0456 mole, 2.5 equivalents) in spectral grade benzene (Note 1, 350 ml) is purged with argon for 45 minutes and irradiated with a 450-watt Hanovia medium-pressure mercury arc lamp in a Pyrex immersion well as a slow stream of argon is passed into the solution. Reaction is complete after 3.25 hours and the crystalline suspension is evaporated to 50 ml on a rotary evaporator. Suction filtration and hot benzene wash (2 × 30 ml) gives analytically pure 1,2-dihydronaphtho[2,1-d]-3*H*-indano-[2,1-f]-*N*-phenylsuccinimido[4,3-b]-7α-thia-2β,3β,5α, 6β-tetrahydrobicyclo-[2.2.1]hepta-2,5-diene (6.56 g, 80.8%), m.p. 264°. Concentration of the combined filtrate and washings gives additional crystalline product (0.29 g, 3.6%, total yield 84.4%) (Note 4). Sublimation of the mother liquor gives recovered *N*-phenylmaleimide (1,3 g).

* Department of Chemistry, Cornell University, Ithaca, New York 14853.

2. Notes

1. Materials were obtained from Eastman Kodak Co. and used as received.

2. For related examples see ref. 2.

3. *N*-phenylmaleimide was prepared by the method of Cava et al.[3] and sublimed (80°, ~0.2 mm) before use.

4. High resolution electron impact mass spectrum, *m/e* (calculated for $C_{29}H_{23}NO_2S$, 449.1457) 449.1450; ir(Nujol): 5.84 (s).

References

1. A. G. Schultz and M. B. DeTar, this volume, p. 119.
2. A. G. Schultz and M. B. DeTar, *J. Am. Chem. Soc.*, **96**, 296 (1974) and A. G. Schultz and M. B. DeTar, *ibid.*, submitted for publication.
3. M. P. Cava, A. A. Deana, K. Muth, and M. J. Mitchell, *Org. Syn.*, **41**, 93 (1961).

cis-DIMETHOXY-1,2-DIOXETANE

$$\underset{H_3CO}{\overset{H}{}}\!\!>C{=}C<\!\!\underset{OCH_3}{\overset{H}{}} \xrightarrow[\text{sensitizer}]{h\nu,\ O_2,\ -78} \text{cis-3,4-dimethoxy-1,2-dioxetane}$$

Submitted by A. P. SCHAAP, A. L. THAYER, and K. KEES*
Checked by T. D. ROBERTS and T. WOOLDRIDGE†

1. Procedure (Note 1)

The photooxidation of *cis*-dimethoxyethylene (Note 2) is conducted in a tubular Pyrex vessel, 15 cm long and 2-cm I.D., with a sintered glass disk bottom 12 cm from the top. Oxygen is introduced into the reaction vessel through this glass disk. The portion of the glass tube below the disk is constricted to 5 mm and connected *via* a three-way stopcock to an oxygen source or vacuum pump as required. The top of the vessel is capped with a rubber septum through which a syringe needle is passed. The syringe needle serves as an oxygen vent and can be used to withdraw aliquots of the reaction solution. The apparatus is held in a Dry Ice-acetone bath at −78° contained in a Dewar with a Pyrex window. A solution of *cis*-dimethoxyethylene (0.11 g, 1.25 mmoles) and tetraphenylporphin (1.5 mg, 0.0024 mmole) (Note 3) in 15 ml of fluorotrichloromethane (Note 4) is irradiated with a 500-watt tungsten-iodide lamp for 30 minutes (Note 5). Oxygen ebullition through the sintered glass disk is maintained during the reaction. White crystals of the dioxetane begin to form in 20–25 minutes and soon fill the vessel (Note 6). The crystalline dioxetane is isolated at −78° by vacuum filtration of the reaction solution through the sintered glass disk (Note 7). The dioxetane is recrystallized once in the reaction vessel from 1 ml of $CFCl_3$.[1] During the recrystallization a small flow of oxygen is maintained through the sintered glass disk to minimize the loss of solution through the disk. The solvent in which the dioxetane is to be dissolved is then added to the crystalline dioxetane at −78° (Note 8). This mixture is allowed to warm to ambient temperature and results in a solution of the dioxetane which can be handled easily. The yield of the dioxetane as determined by iodometric

* Wayne State University, Detroit, Michigan 48202.
† University of Arkansas, Fayetteville, Arkansas 72701.

titration[2] or analysis by nmr with an added internal standard is 45–50%.[1] Thermal decomposition of this dioxetane at 60° gives methyl formate quantitatively. In the presence of a fluorescent hydrocarbon such as 9,10-dibromoanthracene the decomposition is attended with chemiluminescence.[2,3] The nmr spectrum of *cis*-dimethoxy-1,2-dioxetane in $CFCl_3$ exhibits singlets at δ 5.83 (1*H*) and δ 3.63 (3*H*); m.p. 17–18°.

2. Notes

1. The procedure is that of Schaap and Bartlett.[2–4] Mazur and Foote have reported the photooxidative preparation of tetramethoxy-1,2-dioxetane.[5]

2. *cis*-Dimethoxyethylene is prepared by the dealcoholation of 1,1,2-trimethoxyethane. The procedure is adapted from the work of McElvain and Stammer.[6] The material used in the photooxidation was purified by preparative gas chromatography.[3] Chloracetaldehyde dimethyl acetal (Aldrich) is refluxed for 24 hours with one equivalent of sodium methoxide in methanol. The filtered solution is distilled to give 1,1,2-trimethoxyethane (b.p. 123–124°) in 67% yield. The 1,1,2-trimethoxyethane (71 g) is passed through activated alumina (25 g, Alcoa F 8–10) under N_2 at 170°. The products are collected in a Dry Ice-cooled receiver and the methanol removed by distillation. Analysis of the residue by nmr indicates a mixture of approximately 50% *cis*-dimethoxyethylene, 4% *trans*-dimethoxyethylene, and 46% 1,1,2-trimethoxyethane. Preparative gas chromatography (12 ft-20% Carbowax 20 M on Chromosorb P at 98°) of this mixture affords the *cis*-dimethoxyethylene. An alternative synthesis of this alkene has been reported by Shostakovskii et al.[7]

3. Tetraphenylporphin can be purchased from Aldrich Chemical Co., Milwaukee, Wisconsin.

4. The fluorotrichloromethane (Freon 11) obtained from Virginia Chemicals was dried over 3A molecular sieves.

5. The progress of the reaction can be monitored by nmr, following the disappearance of the singlet at δ 5.13 (olefin) and the appearance of the singlet at δ 5.83 (dioxetane). Prolonged irradiation of the reaction solution results in photochemical decomposition of dioxetane to methyl formate.[1]

6. It has been observed that crystallization must sometimes be induced by scratching the inner wall of the reaction vessel.

7. A minor explosion has occurred while handling the pure dioxetane (a liquid) at ambient temperature. In solution the dioxetane is well behaved and decomposes smoothly when heated.

8. Solvents in which this dioxetane has been investigated include benzene, acetone, and methylene chloride.

3. Methods of Preparation

Trimethyl and tetramethyl-1,2-dioxetane have been prepared by base-catalyzed elimination of HBr from the corresponding bromohydroperoxides.[8]

4. Merits of Preparation

The procedure described above has the advantage of yielding a single isomer in high purity from a relatively simple reaction. It appears, however, that as a preparative method for 1,2-dioxetanes the procedure will be synthetically useful only with very reactive olefins such as vinylene diethers.

References

1. T. Wilson and A. P. Schaap, *J. Am. Chem. Soc.*, **93**, 4126 (1971).
2. P. D. Bartlett and A. P. Schaap, *J. Am. Chem. Soc.*, **92**, 3223 (1970).
3. A. P. Schaap, *Tetrahedron Lett.*, 1757 (1971).
4. A. P. Schaap and N. Tontapanish, *Chem. Commun.*, 490 (1972).
5. S. Mazur and C. S. Foote, *J. Am. Chem. Soc.*, **92**, 3225 (1970).
6. S. M. McElvain and C. H. Stammer, *J. Am. Chem. Soc.*, **73**, 915 (1951).
7. M. F. Shostakovskii, N. V. Kusnetsov, and C. Yang, *Isvest. Akad. Nauk S.S.S.R., Otdel. Khim. Nauk*, 1685–1688 (1961); *Chem. Abstr.*, **56**, 5808d (1962).
8. K. R. Kopecky and C. Mumford, *Can. J. Chem.*, **47**, 709 (1969).

3,3-DIMETHYL-2,4-DIPHENYLTRICYCLO[3.2.0.0^{2,4}]HEPT-6-ENE

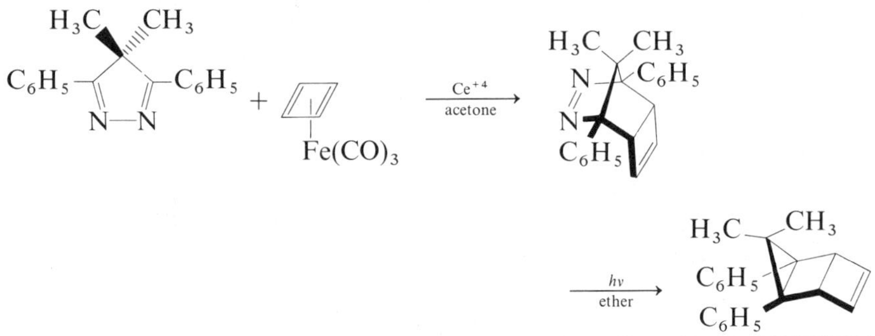

Submitted by L. A. PAQUETTE and L. M. LEICHTER*
Checked by T. D. ROBERTS and C. T. ROBERTS†

1. Procedure

A. endo-7,8-Diaza-9,9-dimethyl-1,6-diphenyltricyclo[4.2.1.0^{2,5}]*nona-3,7-diene.* A solution of 3,5-diphenyl-4,4-dimethylisopyrazole (9.92 g, 0.04 mole) (Note 1) and cyclobutadieneiron tricarbonyl (7.68 g, 0.04 mole) (Note 2) in 700 ml of reagent grade acetone is blanketed with nitrogen and cooled to 0°. During 1.0–1.5 hours ceric ammonium nitrate (109.6 g, 0.20 mole) is added in portions. Twenty minutes after completion of the addition the resultant slurry is poured into anhydrous ether (1 liter) and the precipitated cerium salts are removed by filtration. The filtrate is washed with water and dried. Careful removal of the solvent is followed by chromatography of the crude product on silica gel. Elution with ether-hexane (1:9) and recrystallization from hexane gives 6.6–7.8 g (55–65%) of the azo compound as white crystals, m.p. 153–154° dec. (Note 3).

B. 3,3-Dimethyl-2,4-diphenyltricyclo[3.2.0.0^{2,4}]*hept-6-ene.* A magnetically stirred solution of the azo compound (10.0 g, 0.033 mole) from the previous step in 250 ml of absolute ether is irradiated with a 200-watt Hanovia mercury arc in a water-cooled immersion apparatus (Pyrex). The progress of the reaction, followed by thin-layer chromatography, is usually complete in 3 hours. Careful removal of the solvent affords 9.0 g (100%) of the pure hydrocarbon, m.p. 65–67°, on recrystallization from ethanol (Note 4).

* The Ohio State University, Columbus, Ohio 43210.
† University of Arkansas, Fayetteville, Arkansas 72701.

2. Notes

1. This isopyrazole may be prepared by reaction of the appropriate 1,3-dione with hydrazine according to the procedure of Evnin and coworkers.[1]
2. The cyclobutadieneiron tricarbonyl is available by methods described elsewhere.[2,3]
3. The nmr spectrum (in $CDCl_3$) shows the expected aryl proton multiplet of area 10 at δ 7.32–7.80, an olefinic proton singlet of area 2 at 6.05, a two-proton allyl singlet at 3.88, and methyl singlets at 1.04 and 0.28.
4. The nmr spectrum (in $CDCl_3$) displays a multiplet centered at δ 7.22 (10*H*, aryl), doublets each of area 2 at 6.41 (J = 1.5 Hz, olefinic) and 3.36 (J = 1.5 Hz, allyl), and methyl singlets at 1.44 and 0.88.

3. Methods of Preparation

The procedure described here is based on that originally published by the submitters.[4] It represents a more versatile method of preparing tricyclo[3.2.0.0^{2,4}]hept-6-enes than that involving carbene or carbenoid cycloaddition to hexamethyl (Dewar benzene).[5,6]

References

1. A. B. Evnin, D. R. Arnold, L. A. Karnischky, and E. Strom, *J. Am. Chem. Soc.*, **92**, 6218 (1970).
2. L. A. Paquette and L. D. Wise, *J. Am. Chem. Soc.*, **89**, 6659 (1967).
3. R. Pettit and J. Henery, *Org. Synth.*, **50**, 21 (1970).
4. L. A. Paquette and L. M. Leichter, *J. Am. Chem. Soc.*, **92**, 1765 (1970), **93**, 5128 (1971).
5. E. Müller and H. Kessler, *Tetrahedron Lett.*, 3037 (1968).
6. H. Prinzbach and E. Druckrey, *Tetrahedron Lett.*, 4285 (1968).

3,4-DIMETHYLTRICYCLO[4.4.0.0^{2,8}]DEC-3-ENE-7,10-DIONE

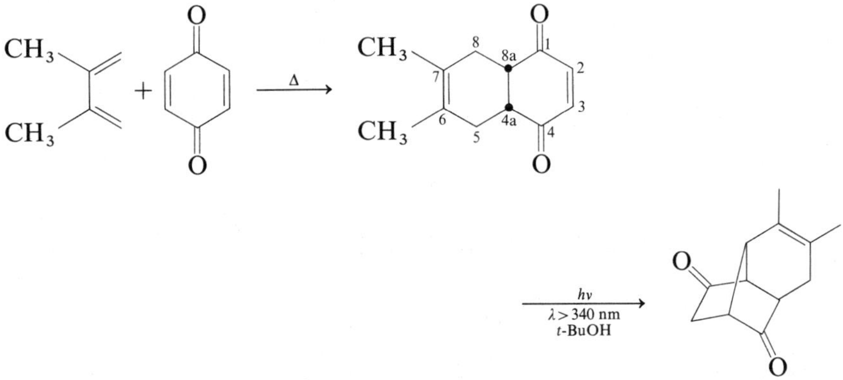

Submitted by JOHN R. SCHEFFER and KULDIP S. BHANDARI*
Checked by J. CORNELISSE and J. N. M. BATIST†

1. Procedure

A. 6,7-Dimethyl-4aβ,5,8,8aβ-tetrahydro-1,4-naphthoquinone. A suspension of *p*-benzoquinone (5.0 g, 0.046 mole) (Note 1) in 2,3-dimethylbutadiene (10 g, 0.12 mole) (Note 2) is heated at 65° for 2 hours. The excess diene is removed by rotary evaporation *in vacuo* to yield a thick slurry which crystallizes. Recrystallization from ethanol gives 6.62 g (75%) of yellow needles of 6,7-dimethyl-4αβ,5,8,8αβ-tetrahydro-1,4-naphthoquinone, m.p. 115–116° (lit.[1] 115–117°).

B. 3,4-Dimethyltricyclo[4.4.0.0^{2,8}]*dec-3-ene-7,10-dione.* The Diels-Alder adduct prepared above (1.0 g, 5.3 mmoles) in 400 ml of 80:20 (v/v) *tert*-butyl alcohol-benzene (Note 3) is placed in a cylindrical Pyrex flask fitted with a magnetic stirring bar, a fritted inert gas inlet disc, and a stopcock-regulated sampling device. After thorough deoxygenation with high purity nitrogen (2 hours) the solution is irradiated *externally* from a distance of 1 ft, using a 275 W Westinghouse uv sun lamp (Note 4) with a 16.5 × 16.4 cm sheet of Corning No. 7380 glass (transmitting $\lambda \geq 340$ nm) interposed between the lamp and the reaction vessel (Note 5). The reaction may be conveniently followed by monitoring the decrease in the uv absorption of the starting material at 350–370 nm (Note 6). The reaction is complete

* University of British Columbia, Vancouver 8, Canada.
† Gorlaeus Laboratoria, Rijksuniversiteit, Leiden, The Netherlands.

when no further change is observed in the uv spectrum in this region (30–35 hours). Rotary evaporation of the solvent *in vacuo* gives a thick oil which is purified by short-path distillation (88–90°, 0.05 mm) in a Kugelrohr apparatus (Note 7) to give 0.8 g (80%) of partly crystalline 3,4-dimethyltricyclo[4.4.0.$0^{2,8}$]dec-3-ene-7,10-dione. Recrystallization from ether-hexane mixtures gives analytically pure material, m.p. 84–85° (needles) (Note 8).

2. Notes

1. Eastman Organic Chemicals Practical Grade benzoquinone was used.
2. 2,3-Dimethyl-1,3-butadiene was obtained from the Aldrich Chemical Co. It can be made according to *Org. Synth. Collect. Vol. 3*, p. 312.
3. The mixed solvent system was used to prevent freezing of the *tert*-butyl alcohol solution. The *tert*-butyl alcohol (Fisher Certified Reagent) was used as supplied; benzene (Fisher Certified A.C.S.) was distilled before use.
4. This lamp, commonly used for suntanning, is readily available in most drug stores in a package form that includes a convenient lamp stand. The checkers used a Rayonet Photochemical Reactor RPR-208 with 8 3500 Å lamps. The irradiation took 9 hours.
5. This filter was obtained from Corning Glass Works, Corning, New York, 14830. C.S. No. 0-52.
6. The concentration of the photolysis solution is such that aliquots can be analyzed directly in this wavelength region without dilution by using 1-cm cells.
7. Kugelrohr Distillation Oven-Rinco Instrument Co., 503 South Prairie Street, P. O. Box 177, Greenville, Illinois, 62246.
8. The checkers isolated the product by column chromatography (Kieselgel H nach Stahl, Type 60, Merck) with a mixture of hexane and chloroform whose ratio changed from 50:50 to 30:60.

3. Discussion

The photochemical reaction described appears to be general for *p*-benzoquinone-open-chain 1,3-diene Diels-Alder adducts which have no substituents at the bridgehead (4a and 8a) or *peri* (C_5 and C_8) positions.[2] The mechanism of the reaction has been shown[2] to involve (a) initial β-hydrogen atom abstraction by oxygen, (b) C_3—C_9 bonding of the diradical thus formed, and (c) ketonization. Other products are formed when pure benzene is used as the photolysis solvent.[2]

The procedure is by far the simplest entry into the tricyclo[4.4.0.$0^{2,8}$]decane ring system found in a number of naturally occurring sesquiterpenes; for example, copacamphor[3] and sativene[4] and their derivatives.

References

1. A. Mandelbaum and M. Cais, *J. Org. Chem.*, **27**, 2243 (1962).
2. J. R. Scheffer, K. S. Bhandari, R. E. Gayler, and R. H. Wiekenkamp, *J. Am. Chem. Soc.*, **94**, 285 (1972).
3. M. Kolbe-Haugwitz and L. Westfelt, *Acta Chem. Scand.*, **24**, 1623 (1970) and references cited therein.
4. P. de Mayo and R. E. Williams, *J. Am. Chem. Soc.*, **87**, 3275 (1965).

1,5-DIPHENYLTRICYCLO[2.1.0.0^{2,5}]PENTAN-3-ONE, METHYL 2,3-DIPHENYLCYCLOPROP-2-ENYLACETATE, AND ITS DIMER

$R = C_6H_5$

Submitted by N. NAKATSUKA and S. MASAMUNE*
Checked by T. D. ROBERTS and T. WOOLDRIDGE†

1. Procedure

To a suspension of 2,3-diphenylcycloprop-2-enylcarboxylic acid (5 g, 0.021 mole)[1] (Note 1) in 25 ml of dry benzene is added *ca.* 5 ml of oxalyl chloride. Within 2 hours at room temperature, with occasional shaking (Note 2), the evolution of gas ceases and the resulting clear solution is evaporated below 40° to afford a crystalline residue which is dissolved in *ca.* 20 ml of dry benzene; the solvent is then evaporated to remove traces of oxalyl chloride. A solution of the crystalline acid chloride in 60 ml of dry benzene is added dropwise to 150 ml of cold (0°) stirred ethereal diazomethane prepared from *N*-nitrosomethylurea (15.4 g, 0.15 mole) and 45 ml of 50% potassium hydroxide[2] (Note 3). The mixture is stirred for 6 hours at 0° and the solvent and excess diazomethane are removed on a rotary evaporator below 40°. The crystalline residue is collected on a funnel and washed with 5 ml of cold ether. The diazoketone thus obtained (4.7–5.1 g) is used directly in the photolysis (Note 4).

A solution of 1 g of this diazoketone in 200 ml of methanol (Note 5) is irradiated at room temperature with a Hanovia 450-watt Type L mercury lamp, using a Pyrex filter. Evolution of nitrogen subsides in approximately 45 minutes and the crystalline material deposited is filtered.

Five batches of the above photolysis provides 0.8 to 1.1 g of precipitate which is recrystallized from hot dioxane (17–25 ml) to provide 0.5 ~ 0.9 g of the dimer, m.p. 236–237°.

* University of Alberta, Edmonton, Alberta, Canada.
† University of Arkansas, Fayetteville, Arkansas 72701.

The combined filtrates, after evaporation on a rotary evaporator, provide 4.0 to 4.5 g of brown residue which is chromatographed over 250 g (6 × 18 cm) silicic acid (Mallinckrodt, 100 mesh, analytical reagent) using $CHCl_3$ (0.75% ethanol) as an eluent.

Fraction	Volume (ml)	Weight (mg)
1	250	2
2	250	39
3	250	218
4	250	1334
5	125	68
6	125	262
7	250	451
8	250	170
9	250	10

Fraction 4 is recrystallized from methanol to provide 1.1 g (21%) of methyl 2,3-diphenylcycloprop-2-enylacetate, m.p. 99–100°. Fractions 6 to 8 are combined and recrystallized from ether (Note 6) to give two crops, 0.55 g and 0.15 g (total, 12%), of 4,5-diphenyltricyclo[2.1.0.0^{2,5}]pentan-2-one, both melting at 141.5-142.5° with decomposition.

2. Notes

1. A sample melting above 205° should be employed.
2. The mixture may be warmed to 40° to initiate quick gas evolution at the beginning.
3. Distillation of the ethereal diazomethane is not necessary. See Note 3 of ref. 2. If *N*-nitrosomethylurea stabilized with acetic acid is employed, the amount of KOH is increased accordingly.
4. Do not purify the diazoketone further.[3]
5. When the diazoketone is contaminated with insoluble polymethylene, the solution is filtered.
6. All the material is dissolved in a rather large amount of ether and the solution is concentrated rapidly at atmospheric pressure until crystals begin to form.

3. Methods of Preparation

When tetrahydrofuran is used as solvent in the above photolysis,[3] the tricycloketone is the only isolable product in a yield comparable to that described above. The chromatographic separation is somewhat simpler. Copper-catalyzed reaction of the diazoketone in refluxing benzene affords a trace amount of the tricycloketone in, at most, 1% yield. A similar observation is made with the corresponding dimethyl derivative.[4] The base-catalyzed decomposition of the tosylhydrazone derived from 1,3-dimethylcyclobut-2-enylcarboxaldehyde provides 2,4-dimethyltricyclo[2.1.0.0^{2,5}]pentane.[5] Treatment of the diazoketone (7 g) in methanol (180 ml) with 1 g of silver benzoate and 10 ml of triethylamine at 40° for 5 hours provided 3.5 to 4.0 g of homologated methyl ester.

4. Merits of Preparation

1,5-Diethyl and 1,5-dipropyltricyclo[2.1.0.0^{2,5}]pentan-3-one are prepared from their corresponding diazoketones in 10 to 15% yield, whereas the copper-catalyzed addition proceeds inefficiently to provide low yields (1%) of the ketones. Thus the photochemical reaction has a definite advantage. In contrast, the preparation of the homologous ketone, substituted tricyclo-[3.1.0.0^{2,6}]hexan-4-one, is best achieved by the copper reaction of the diazoketone derived from 2,3-diphenylcycloprop-2-enyl acetic acid.[6] Photolysis of the latter compound undergoes a Wolff-type rearrangement.[7] 1,5-Disubstituted tricyclo[2.1.0.0^{2,5}]pentan-3-ones have served as a precursor to generate the unique C_5H_5 carbocation with the square pyramidal structure.[8]

References

1. R. Breslow, R. Winter, and M. Battiste, *J. Org. Chem.*, **24**, 415 (1959).
2. F. Arndt, *Org. Synth., Collect. Vol. II*, 165 (1941).
3. J. Trotter, C. S. Gibbons, N. Nakatsuka, and S. Masamune, *J. Am. Chem. Soc.*, **89**, 2793 (1967). S. Masamune, *ibid.*, **86**, 735 (1964).
4. W. von E. Doering and M. Pomerantz, *Tetrahedron Lett.*, 961 (1964).
5. G. L. Closs and R. B. Larrabee, *ibid.*, 287 (1965).
6. A. S. Monahan, *J. Org. Chem.*, **33**, 1441 (1968). S. Masamune and N. T. Castellucci, *Proc. Chem. Soc.*, 298 (1964).
7. A. S. Monahan and S. Tang, *J. Org. Chem.*, **33**, 1445 (1968). S. Masamune and K. Fukumoto, *Tetrahedron Lett.*, 4647 (1965).
8. S. Masamune, M. Sakai, H. Ona, and A. J. Jones, *J. Am. Chem. Soc.*, **94**, 8956 (1972).

1-ETHOXYCARBONYL-(1*H*)-1,2-DIAZEPINE

$$NaN_3 + ClCO_2Et \longrightarrow N_3CO_2Et + NaCl$$

$$N_3CO_2Et + \text{pyridine} \longrightarrow \text{N-ethoxycarbonylimino-pyridinium ylid} + N_2$$

$$\text{N-ethoxycarbonylimino-pyridinium ylid} \xrightarrow{h\nu} \text{1-ethoxycarbonyl-(1}H\text{)-1,2-diazepine}$$

Submitted by M. NASTASI, E. S. SCHILLING and J. STREITH*

1. Procedure

Caution! The first two steps (A) and (B) should be carried out in an efficient and strong-shielded hood to avoid exposure to ethyl azidoformate vapors and possible explosion hazards.

A. Ethyl azidoformate. Ethyl chloroformate (32.4 g; 0.3 mole) is added dropwise over a period of 0.5 hour to a stirred solution of sodium azide (32.5 g; 0.5 mole) (Note 1) in 100 ml of water. The mixture is stirred at room temperature for an additional period of 15 hours and then transferred into a 500-ml separatory funnel. After removal of the water solution the colorless organic layer is dried over sodium sulfate; yield, 83%.

B. N-ethoxycarbonylimino-pyridinium ylid (Note 2). A 500-ml three-necked, round-bottomed flask equipped with a thermometer, a condenser, and a funnel is charged with pyridine (100 g; 1.27 moles) (Note 3). The pyridine is brought to reflux and ethyl azidoformate (28.7 g; 0.25 mole) is added dropwise through the funnel over a period of 1 hour. The mixture is further stirred under reflux for about 40 hours; excess pyridine is then distilled off under vacuum in a rotary evaporator. The resulting darkbrown crystalline solid is dissolved in 300 ml of methanol and the solution treated three times under reflux (Note 4) with charcoal; after filtration a light yellow solution is obtained. Following removal of the solvent under vacuum, the yellow solid residue is recrystallized twice from benzene-hexane (7:3) to afford colorless needles (16.7 g; yield: 40%); m.p. 109°C; uv $(C_6H_6)\lambda_{max}$: 342 nm (ε: 12,000).

* Ecole Supérieure de Chimie de Mulhouse, Mulhouse Cedex, France.

C. 1-Ethoxycarbonyl-(1H)-1,2-diazepine[2,3]. A solution of *N*-ethoxycarbonylimino-pyridinium ylid (8 g; 0.05 mole) in 4 l. benzene (Note 5) is irradiated under a nitrogen atmosphere by means of a Philips HPK 125 high-pressure mercury lamp, through a Pyrex filter in a 4-litre water-cooled immersion-well reactor. The solution turns progressively orange. The rate of conversion is monitored spectrophotometrically by following the disappearance of the 342-nm absorption band. In about 10 hours all the starting material disappears. The solvent is removed under a vacuum in a rotary evaporator and the resulting oily red-orange residue chromatographed over a 300-g silica gel column. Elution with cyclohexane: ethylacetate (6:4) affords 7.6 g of a viscous red-orange oil which may be further purified by vacuum distillation (yield: 95%); b.p. 51°C (under 0.1 mm Hg); uv (MeOH) λ_{max} 217 nm (ε: 11,200) and 362 nm (ε: 280); n_D^{22} = 1.5302; nmr ($CHCl_3$-d) τ 2.6 (1*H*; q; *J* = 3.5 and 1.2 Hz), τ 3.42 (1*H*; m; *J* = 11.0 and 1.2 Hz), τ 3.75 (1*H*; m; *J* = 3.5 Hz), τ 3.82 (1*H*; m; *J* = 1.0 and 7.5 Hz), τ 4.25 (1*H*; m; *J* = 1.5 and 5 Hz), τ 5.68 (2*H*; q; *J* = 7.0 Hz) and τ 8.65 (3*H*; t; *J* = 7.0 Hz).

2. Notes

1. Ethyl chloroformate and sodium azide were purchased from the FLUKA AG Co. and used without any further purification.
2. This procedure is essentially the one described by Hafner et al[1].
3. Pyridine was purchased from Rhone-Poulenc Co., RP purity grade, and dried over sodium hydroxide pellets before use.
4. Charcoal was purchased from FLUKA AG Co. and is of "PURUM" Grade.
5. Commerical benzene was purchased from Rhone-Poulenc Co., RP purity grade, and used without any further purification.

References

1. K. Hafner, D. Zinser, and K. L. Moritz, *Tetrahedron Lett.*, 1733 (1964).
2. J. Streith and J. M. Cassal, *Angew. Chem.*, **80**, 117 (1968); *Angew. Chem. Int. Ed.*, **7**, 129 (1968); J. Streith and J. M. Cassal, *Tetrahedron Lett.*, 4541 (1968); J. Streith and J. M. Cassal, *Bull. Soc. Chim. France*, 2175 (1969).
3. For a recent review of this topic see J. M. Cassal, A. Frankowski, J. P. Luttringer, M. Nastasi, J. Streith, G. Taurand, and B. Willig, Lectures in Heterocyclic Chemistry, Raymond N. Castle, Ed., *J. Heterocycl. Chem.*, Supplementary Issue to Vol. 9.

1,3,4,5,6,6-HEXAMETHYLBICYCLO[3.1.0]HEX-3-EN-2-ONE AND 2,3,4,5,6-PENTAMETHYL-*N*-CYCLOHEXYLHEPTA-3,5-DIENAMIDE

$h\nu$, CH_3OH

$h\nu$, $C_6H_{11}NH_2$-Et_2O

$CH_3—C(CH_3)═C(CH_3)—C(CH_3)═C(CH_3)—CH(CH_3)—C(═O)—NH—C_6H_{11}$

Submitted by DAVID A. DICKINSON, THOMAS A. HARDY, and HAROLD HART*
Checked by JOSE A. ORS†

1. Procedure

A. 1,3,4,5,6,6-Hexamethylbicyclo[3.1.0]hex-3-en-2-one. A solution of 2,3,4,5,6,6-hexamethylcyclohexa-2,4-dienone (17.8 g, 0.100 mole) (Note 1) in 350 ml of reagent grade methanol is deoxygenated by passing a stream of nitrogen through the solution for 15 minutes. The solution is irradiated under nitrogen with a pyrex-filtered 450-watt Hanovia medium-pressure mercury lamp in an immersion apparatus for 60 minutes (Note 2). When the reaction as monitored by vpc, is nearly complete (Note 3), the methanol is removed by rotary evaporation and the photoketone is distilled through a 10-cm Vigreux column (62–66° at 0.5 torr) to yield 15.0 g (Note 4) of a yellow oil which solidifies on standing (m.p. 46.5–48.5°). Further purification can be effected by sublimation or recrystallization from methanol-water (m.p. 48–49°).

B. 2,3,4,5,6-Pentamethyl-N-cyclohexylhepta-3,5-dienamide. In the same apparatus described above a solution of 2,3,4,5,6,6-hexamethylcyclohexa-2,4-dienone (17.8 g, 0.100 mole) and freshly distilled cyclohexylamine (57.2 ml, 0.500 mole) (Note 5) in 300 ml of anhydrous ethyl ether is irradiated for 90 minutes. Analysis of the photolysate by vpc (Note 6) indicates complete reaction of starting material. The excess amine is removed by washing with

* Michigan State University, East Lansing, Michigan 48824.
† IBM Thomas J. Watson Research Center, Yorktown Heights, New York 10598.

500 ml of 5% aqueous hydrochloric acid. The solvent is removed by rotary evaporation and the amide is recrystallized from methanol-water to yield 23.8 g (0.086 mole) of white plates melting from 76–77.5° (Note 7).

2. Notes

1. 2,3,4,5,6,6-Hexamethylcyclohexa-2,4-dienone is prepared in good yield by oxidation of hexamethylbenzene.[1]
2. It is important to stop the irradiation as soon as the starting dienone has disappeared, because prolonged irradiation of the photoketone gives 2,3,3,4,5,6-hexamethyl-6-methoxycyclohexa-1,4-dien-1-ol.[2]
3. On a 5 ft × 1/4 in. 5% SE-30 column at 150° with 20 ml/min of the He carrier gas the retention times of the starting dienone and photoproduct are 7 and 4 minutes, respectively.
4. Analysis by vpc shows the photoketone to be *ca.* 95% pure. Isolated yields range from 60–85%.
5. If excess amine is not used, some of the intermediate ketene is not trapped and cyclizes to give the product described in Part A.
6. On a 5 ft × 1/4 in. 5% SE-30 column at 200° with 20 ml/min of He carrier gas the amide has a retention time of 6 minutes.
7. The amide was 98–99% pure, obtained in yields ranging from 70–90%. The stereochemistry at the 3,4-double bond is not known; if the initially formed ketene is trapped without isomerization it should be *Z*.

3. Methods of Preparation

The procedures described here are taken from the literature[1,3] but have been scaled up and improved.

4. Merits of the Preparation

The actual photoproduct in each reaction is the ketene $(CH_3)_2C{=}C(CH_3)$-$C(CH_3){=}C(CH_3)C(CH_3){=}C{=}O$.[3,4] Methanol is not sufficiently nucleophilic to trap the ketene, which in this solvent cyclizes efficiently to the bicyclo[3.1.0]hexenone.

The bicyclic ketone, in acid, provided the first example of a [1,4]sigmatropic shift in carbonium ions.[5] Other highly substituted 2,4-cyclohexadienones also give bicyclo[3.1.0]hex-3-en-2-ones on irradiation.[1,3,6]

The preparation in Part B is more typical for 2,4-cyclohexadienones and is often the only type of reaction observed when less substituted dienones are irradiated in nucleophilic solvents.[4,7] It provides an excellent general method for preparing diene-acids and their derivatives.

References

1. H. Hart, P. M. Collins and A. J. Waring, *J. Am. Chem. Soc.*, **88**, 1005 (1966).
2. H. Hart and D. W. Swatton, *J. Am. Chem. Soc.*, **89**, 1874 (1967).
3. J. Griffiths and H. Hart, *J. Am. Chem. Soc.*, **90**, 3297 (1968).
4. For a review see G. Quinkert, *Angew. Chem., Int. Ed. Engl.*, **12**, 1072 (1972).
5. D. W. Swatton and H. Hart, *J. Am. Chem. Soc.*, **89**, 5075 (1967).
6. P. M. Collins and H. Hart, *J. Chem. Soc. C.*, 895 (1967); H. Hart and R. K. Murray, Jr., *J. Org. Chem.*, **32**, 2448 (1967); H. Hart and D. C. Lankin, *J. Org. Chem.*, **33**, 4398, (1968); J. Griffiths and H. Hart, *J. Am. Chem. Soc.*, **90**, 5296 (1968); H. Perst and K. Dimroth, *Tetrahedron*, **24**, 5385 (1968); M. R. Morris and A. J. Waring, *Chem. Commun.*, 526 (1969); H. Hart and R. K. Murray, Jr., *J. Org. Chem.*, **35**, 1535 (1970); H. Perst and I. Weisemeier, *Tetrahedron Lett.*, 4189 (1970); M. R. Morris and A. J. Waring, *J. Chem. Soc. C.*, 3266, 3269 (1971); A. J. Waring, M. R. Morris, and M. M. Islam, *J. Chem. Soc. C.*, 3274 (1971); H. Hart and R. J. Bastiani, *J. Org. Chem.*, **37**, 4018 (1972).
7. D. H. R. Barton and G. Quinkert, *J. Chem. Soc.*, 1 (1960).

17β-HYDROXY-5α,19-CYCLO-A-NOR-10α-ANDROSTAN-3-ONE

OH OH

$h\nu$ $\longrightarrow$

O CH_2 O

Submitted by J. R. WILLIAMS* and H. ZIFFER†
Checked by T. D. ROBERTS‡

1. Procedure

In a vessel equipped with a nitrogen inlet, reflux condenser, and an insert containing a pyrex envelope a solution of 17β-hydroxy-5(10)-estrene-3-one (0.5 g) (Note 1) in 150 ml of *tert*-butanol is photolyzed under a stream of dry nitrogen with a Hanovia type L 450-watt medium pressure mercury vapor lamp. The reaction is monitored by vapor-phase chromatography (Note 2). The reaction is stopped after about 30 hours, although some starting material may remain. Longer irradiations lead to some destruction of the photoproduct and lower yields. The solvent is evaporated *in vacuo* to yield a crystalline residue. The product from three runs is combined (1.5 g) and chromatographed over silica gel (Merck H, 55 g) (Note 3). Initial elution with ethyl acetate: hexane 1:4 (700 ml) affords 380 mg of at least three minor products. Further elution with the same solvent yields 150 mg of starting material, followed by 830 mg (55%) of 17β-hydroxy-5α,19-cyclo-A-nor-10α-androstan-3-one which crystallizes from ethyl acetate as colorless prisms m.p. 162–163.5°.[1]

2. Notes

1. 17β-hydroxy-5(10)-estrene-3-one (m.p. 197.5–199.5° dec. vac.)[2,3] was prepared by Birch reduction followed by oxalic acid hydrolysis of 3,17β-hydroxyestra-1,3,5(10)-triene-3-methyl ether. The method of Wilds and Nelson,[2] as modified by Dryden et al.,[3] was used.

2. Vapor-phase chromatography was carried out on a 6 ft glass column, 3-mm I.D., with 3% OV-17 on gas chrom Q (60/180 mesh). The column

* Temple University, Philadelphia, Pennsylvania 19122.
† N.I.A.M.D., National Institutes of Health, Bethesda, Maryland 20014.
‡ University of Arkansas, Fayetteville, Arkansas 72701.

temperature was 210°C. The relative retention time of the photoproduct to the starting ketone is 0.81.

3. A series of small batches was found to give the greatest yield of photoproduct. Because of the low solubility of the starting material, it is necessary to heat the *t*-butanol to form a saturated solution. Because of the similarity in properties between the starting material and photoproduct, the chromatography has to be done very carefully for optimum yield. Merck's silica gel H, suitable for thin layer chromatography, was found to give the best separations. However, it also requires the use of pressure to force the eluent down the column.

3. Methods of Preparation

The only reported synthesis of this cyclopropyl ketone utilizes this procedure.

4. Merits of the Preparation

This procedure describes a practical method for the synthesis of 17β-hydroxy-5α,19-cyclo-A-nor-10α-androstan-3-one and gives a ready entry into the A-nor-10α substituted steroids. The stereospecific reaction leads only to the α cyclo bridge.

References

1. J. R. Williams and H. Ziffer, *Tetrahedron*, **24**, 6725 (1968).
2. A. L. Wilds and N. A. Nelson, *J. Am. Chem. Soc.*, **75**, 5366 (1953).
3. H. L. Dryden, G. M. Weber, R. R. Burtner, and J. A. Cella, *J. Org. Chem.*, **26**, 3237 (1961).

3β-HYDROXY-13,17-SECOANDROSTA-5,13-DIEN-17-OIC ACID

Submitted by J. BRUSSEE and H. J. C. JACOBS*
Checked by J. CORNELISSE and J. N. M. BATIST*

1. Procedure

A solution of 3β-hydroxyandrost-5-en-17-one acetate (I) (6 g, 0.018 mole) (Note 1) in 600 ml of methanol (Note 2) is placed in a 750-ml cylindrical vessel fitted with a Pyrex immersion well (Note 3). The solution is stirred magnetically while being irradiated with a 150-watt high-pressure mercury arc (Hanau TQ 150). The temperature is controlled by cooling the reaction flask with an ice bath. Progress of the reaction can be conveniently monitored by GC analysis (Notes 4, 5).

* Gorlaeus Laboratories, Department of Organic Chemistry, University of Leiden, The Netherlands.

The irradiation is stopped when GC analysis shows no further increase of the aldehyde (II) (usually within 5–8 hours). The solvent is then removed at reduced pressure and the resulting colorless liquid is dissolved in a mixture of 100 ml of acetone (Note 6) and 5 ml of water. Cooling to 0° is followed by addition of a cooled solution of 1.5 g of potassium permanganate in 50 ml of acetone (Note 6) and 10 ml of water. The mixture is kept at 0° and stirred for 2 hours. An ice-cold 10% solution of sodium bisulfite (250 ml) is then added, followed by 200 ml of ethyl ether, and sufficient 1 *N* sulfuric acid solution just to dissolve any precipitated manganese dioxide.

The lower layer is removed and washed twice with 200-ml portions of ethyl ether, which are combined with the original upper layer. The combined organic layer is extracted with three 100-ml portions of 5% potassium hydroxide solution. The combined alkaline extract is then washed with 150 ml of ethyl ether and allowed to stand at room temperature for 30 minutes to complete the acetate hydrolysis. The solution is acidified with 2 *N* sulfuric acid and extracted with three 200-ml portions of ethyl ether. The ether solution is washed with water, dried over anhydrous sodium sulfate, and filtered to remove the drying agent. Evaporation of the solvent at reduced pressure then gives 1.8–2.2 g of white crystalline material, which is recrystallized from acetone to yield 1.40–1.75 g (25–32%) of 3β-hydroxy-13,17-secoandrosta-5,13-dien-17-oic acid (V), m.p. 171–173°, $[\alpha]_D^{23}$ $-220 \pm 5°$ ($CHCl_3$, *c* 0.3) (Note 7).

2. Notes

1. The steroid was obtained as the alcohol from Laboratorios Julian de Mexico, S.A., and used after acetylation and purification in the usual way.
2. Redistilled practical grade methanol can be used; the water content should not be less than 0.1% to prevent formation of the dimethyl acetal of (II).
3. Duran 50, transmittance 20% at λ 300 nm.
4. A Hewlett-Packard F & M 402 Gas Chromatograph was used; the 1.8-meter U-shaped glass column was packed with 1% OV-17 on GasChrom Q (80–100 mesh).
5. In addition to the aldehyde (II), a few other photoproducts are formed,[1] the most abundant being the methyl ester (III) and the 13α-steroid (IV). The latter is eventually photolyzed into (II) and (III). Relative retention times on GC: (I) 1.00; (II) 0.63; (III) 0.75; (IV) 0.67.
6. Freshly distilled from potassium permanganate.
7. The purity of the product, checked by GC analysis of its methyl ester, exceeds 95%. An analytical sample melts at 174–175° and shows $[\alpha]_D^{23}$ $-223°$.

3. Methods of Preparation

3β-Hydroxy-13,17-secoandrosta-5,13-dien-17-oic acid has also been prepared[2] in a multistep synthesis from Koster-Logemann ketone and by a procedure[3] analogous to the one described above but involving isolation of the aldehyde (II) and use of silver oxide as an oxidant (yield not stated).

4. Merits of the Preparation

The present procedure has the advantage of using an inexpensive, commercially available starting material and yielding the secosteroid in relatively high yield and in a minimal number of synthetic operations, without the necessity of isolating the intermediates. It can be adapted easily to the preparation of various other types of 13,17-secosteroids; for example ring A and/or B aromatic compounds.

The product acid (V) has been used as an intermediate in the total synthesis of steroids with unnatural configuration.[2]

References

1. G. Quinkert, *Angew. Chem.*, **77**, 229 (1965); *Angew. Chem. Intern. Ed. Engl.*, **4**, 211 (1965).
2. J. R. Billeter and K. Miescher, *Helv. Chim. Acta*, **34**, 2053 (1951).
3. G. Quinkert and H. G. Heine, *Tetrahedron Lett.*, 1659 (1963).

7-*cis*-β-IONOL

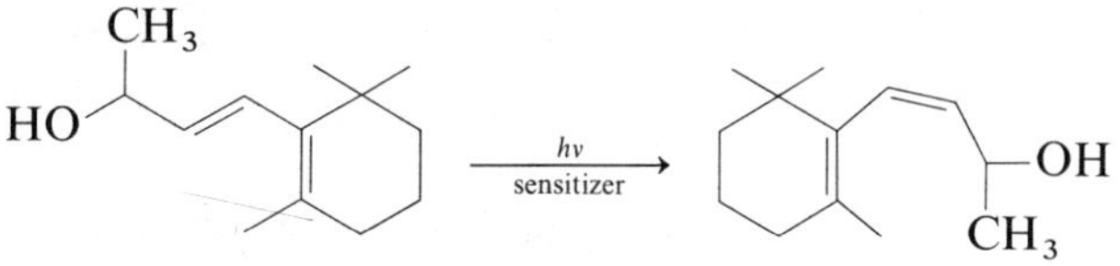

Submitted by V. RAMAMURTHY and R. S. H. LIU*
Checked by T. D. ROBERTS and C. T. ROBERTS†

1. Procedure (Note 1)

A solution of 7-*trans*-β-ionol (10.0 g, 0.051 mole) (Note 2) and β-acetonaphthone (0.4 g) (Note 3) in 100 ml of benzene (Note 4) is added to a water-cooled Pyrex (Note 5) immersion apparatus similar to that shown on p. 14, Volume 1, of this series. Before irradiation the solution is deoxygenated by passing a stream of nitrogen through it for about 2 minutes. While keeping under nitrogen, the solution is irradiated with a 200-watt Hanovia medium-pressure mercury lamp. After approximately 48 hours the reaction is complete (Note 6).

The irradiated solution is transferred to a round-bottomed flask and concentrated by evaporation on a rotary evaporator. The concentrated solution is distilled under vacuum through a 8-in. Vigreux column. The colorless liquid which boils at 91–2°C (1.2 mm) (8.6 g, 86% yield) (Note 7), is 7-*cis*-β-ionol.

2. Notes

1. The same procedure, with some minor modifications such as using a different filtering system, has been used to effect one-way *trans*-to-*cis* isomerization of several other hindered olefins such as hindered styrenes, stilbenes, and β-ionylidene derivatives.[1,2]

2. *trans*-β-Ionol was prepared by sodium borohydride reduction of β-ionone following a procedure similar to that described by Mousseron-Canet.[3] The ketone was obtained from the Eastman-Kodak Co. and used without further purification. Samples obtained from other suppliers are sometimes contaminated with α-ionone.

* University of Hawaii, Honolulu, Hawaii 96822.
† University of Arkansas, Fayetteville, Arkansas 72701.

3. One-way conversion of the *trans* isomer to the *cis* can also be effected by other sensitizers with triplet energy between 55 and 66 kcal/mole. Among these are triphenylene, β-naphthyl phenyl ketone, and α-acetonaphthone.

4. Concentrations of material are not critical. Other inert solvents such as *n*-hexane and methanol can also be used. The checkers used heptane.

5. Direct absorption of light by ionol results in an irreversible side reaction of sigmatropic hydrogen migration. The Pyrex vessel serves to filter off the unwanted 2537 Å light.

6. The progress of the reaction can be conveniently followed by GLPC analyses (5 ft, 3% SE-30 column at 128°) or by nmr spectroscopy. The nmr spectra of both isomers have been published.[4,6]

7. The quantum yield of the reaction, calculated from its decay ratio, is 0.65.[1]

3. Methods of Preparation

7-*cis*-β-Ionol has also been prepared by sodium borohydride reduction of *cis*-ionone.[5,6]

References

1. V. Ramamurthy, Y. Butt, C. Yang, P. Yang, and R. S. H. Liu, *J. Org. Chem.*, **38**, 1247 (1973); V. Ramamurthy, G. Tustin, C. C. Yau, and R. S. H. Liu, *Tetrahedron*, **31**, 193 (1975).
2. C. S. C. Yang and R. S. H. Liu, *Tetrahedron Lett.*, 4811 (1973).
3. M. Mousseron-Canet, M. Mousseron and P. Legendre, *Bull. Soc. Chim. France*, 50 (1964).
4. R. S. H. Liu, *Pure Appl. Chem.*, Supp. I, 335 (1971).
5. Unpublished results of V. Ramamurthy.
6. E. N. Marvell, T. Chadwick, G. Caple, T. Gosink, and G. Zimmer, *J. Org. Chem.*, **37**, 2992 (1972).

5-METHOXY-2,2-DIMETHYL-5-TRIFLUOROMETHYL-4-PHENYL-3-OXAZOLINE

$$\text{2,2-dimethyl-3-phenyl-2}H\text{-azirine} + CF_3C(=O)OCH_3 \xrightarrow{h\nu} \text{5-methoxy-2,2-dimethyl-5-trifluoromethyl-4-phenyl-3-oxazoline}$$

Submitted by P. UEBELHART, P. GILGEN, and H. SCHMID*
Checked by J. CORNELISSE and Mrs. G. M. GORTER-LAROY†

1. Procedure

A solution of 2,2-dimethyl-3-phenyl-2*H*-azirine (2.17 g, 15 mmoles) (Note 1) in 800 ml of benzene (Note 2) is placed in the reaction vessel (Note 3) and a gentle stream of dry nitrogen is passed through it for 30 minutes. After that period methyl trifluoroacetate (2.50 g, 19.5 mmoles) (Note 4) is added and nitrogen is bubbled through for another 5 minutes. The lamp (Note 5) is then inserted in the well and the irradiation is started. The course of the reaction can be followed best by gas-chromatographic analysis (Note 6). After about 3 hours of irradiation the conversion of the starting material is essentially complete.

The solvent is removed on a rotary evaporator under reduced pressure at approximately 30°C. The oily residue crystallizes on standing at 0°C. The crude product is sublimed under reduced pressure (10^{-2} torr). Some oily impurities distil at 35°C and are removed by washing the condenser with ether. Sublimation is then continued at 50°C to give 5-methoxy-2,2-dimethyl-5-trifluoromethyl-4-phenyl-3-oxazoline as colorless needles, m.p. 56–59°C. The yield is 3.20 g (77%). After recrystallization from pentane the pure product melts at 59.5–60.5°C.

2. Notes

1. The preparation of the starting material from isopropyl phenyl ketone is given by Leonhard and Zwanenburg.[1]

* University of Zurich, Zurich, Switzerland.
† Gorlaeus Laboratoria, Rijksuniversiteit, Leiden, The Netherlands.

2. Commercial benzene (analytical grade, Merck) was used without further purification.

3. The apparatus consists of a cylindrical 800-ml irradiation vessel made of Pyrex and equipped with a nitrogen inlet tube that ends at the bottom of the vessel.

4. The commercial (purum, Fluka) methyl trifluoroacetate was distilled from potassium carbonate just before use.

5. The light source was a water-cooled medium-pressure lamp "HANAU TQ 150."

6. Gas-chromatographic analyses were carried out on a "CARLO ERBA" instrument (model GI) equipped with a silicone-coated (XE 60) glass capillary column[2] (22 × 0.35 mm) and hydrogen as carrier.

3. Methods and Merits of Preparation

The photolysis of 2,2-dimethyl-3-phenyl-2*H*-azirine[3] leads first to the 1,3-dipolar species benzonitrile isopropylid which reacts with methyl trifluoroacetate to form the 5-methoxy-2,2-dimethyl-5-trifluoromethyl-4-phenyl- 3-oxazoline. It is the only method reported for the preparation of this heterocyclic. This reaction also provides the first case of an addition of an ester-C=O—bond to a 1,3-dipole.

References

1. N. J. Leonard and B. Zwanenburg, *J. Am. Chem. Soc.*, **89**, 4456 (1967).
2. K. Grob, *Helv. Chim. Acta*, **48**, 1362 (1965), **51**, 718 (1968).
3. P. Claus, Th. Doppler, N. Gakis, M. Georgarakis, H. Giezendanner, P. Gilgen, H. Heimgartner, B. Jackson, M. Märky, N. S. Narasimhan, H. J. Rosenkranz, A. Wunderli, H. -J. Hansen, and H. Schmid, *Pure Appl. Chem.* **33**, 339 (1973). A. Padwa, M. Dharan, J. Smolanoff, and S. I. Wetmore, Jr., *ibid.*, **33**, 269 (1973).

2-METHOXY-1,8-DIPHENYL-1a,2,7,7a-TETRAHYDRO-1,2,7-METHENO-1H-CYCLOPROPA(b)NAPHTHALENE

OCH_3 (1-methoxynaphthalene) + Ph—C≡C—Ph $\xrightarrow{h\nu}$ (product, Ph, Ph, R, R′)

(I) R = OCH_3; R′ = H
(II) R = H; R′ = OCH_3

Submitted by W. H. F. SASSE*
Checked by T. D. ROBERTS†

1. Procedure

The reaction vessel is a cylindrical Pyrex tube (O.D., 8 cm; length excluding joint 25 cm), which is fitted with a 60/50 ground socket carrying a Hanovia quartz immersion well (No. 19454), two side arms with 19/26 sockets, and, at the center of its lower end, a sintered glass plate (diameter 2 cm; porosity No. 2) through which nitrogen can be introduced. One side arm is stoppered; the other carries an efficient reflux condenser whose outlet to the atmosphere carries a nonreturn valve (Note 1). Benzene (50 ml; Note 2) is introduced and the nitrogen flow (Note 3) is adjusted to prevent leakage of benzene through the sintered plate. Next a solution of 1-methoxynaphthalene (50 g; 0.32 mole; Note 4) and diphenylacetylene (14 g; 0.079 mole; Note 5) in benzene (450 ml) is introduced. Without interrupting the flow of nitrogen the lower end of the reactor is lowered into a beaker (5-liters) containing water (*ca.* 3 liters) and fitted with an immersion heater (500 watts). The water in the beaker is heated until the solution in the reactor has boiled for 2–3 minutes. The reactor is then lifted out of the water and the nitrogen flow is increased to compensate for the pressure drop inside the reactor. After 3 minutes more the reactor is lowered again into the beaker and reheated as above. The beaker is then removed, the immersion well is connected to cooling water, and, with a fast stream of nitrogen passing through the reactor, the immersion well is connected to cooling water until its contents are cooled to *ca.* 30°. The gas flow is then reduced to the minimum that prevents leakage through the plate and a Pyrex filter (Hanovia 7740

* Division of Applied Organic Chemistry, C.S.I.R.O., P. O. Box 4331, G.P.O. Melbourne, Victoria 3001, Australia.
† University of Arkansas, Fayetteville, Arkansas 72701.

Pyrex) and a Hanovia 450-watt mercury high-pressure lamp are inserted into the well. After irradiation for 96 hours diphenylacetylene cannot be detected by GLC (Note 6). At this point the mixture can be worked up as described below (Note 7) or a second lot of diphenylacetylene (7 g; solid) can be added through the second side arm. Without further degassing the mixture is irradiated for another 48 hours. The reaction mixture is then freed of benzene (at 50–60°/30–50 mm), treated with boiling light petroleum (500 ml; Note 8), and filtered. The precipitated adduct (I) (27.5–31.5 g; 70–80%) forms a finely divided off-white powder, m.p. 147–149° (Note 9). By crystallization from boiling anhydrous ethanol (70–80 ml per g) adduct (I) is obtained as colorless needles, m.p. 148–150° (recovery 80–90%).

2. Notes

1. A "Quickfit" valve 2NRV was used.
2. Anhydrous benzene of analytical purity was used.
3. High-purity "oxygen-free" nitrogen was used.
4. 1-Methoxynaphthalene (Fluka; purum) was passed through a dry bed of alumina (1.5 × 10 cm) to remove all colored matter.
5. The diphenylacetylene must be free of stilbene.[1] The checker performed the reaction on 1/4th scale with a Hanovia 500-watt lamp. Irradiation time was 60 hours.
6. A stainless steel column 1 m × 2 mm packed with 2% SE-30 on Chomosorb W was used.
7. At this stage the yield of crude product is 20–22 g (75–83%).
8. Light petroleum, b.p. 60–80°, containing less than 1% aromatics was used.
9. The filtrate contains mainly the adduct (I) accompanied by smaller quantities of its 3-isomer and of 4-oxo-1,2-diphenyl-2a,3,4,8b-tetrahydrocyclobuta(a)naphthalene.[2]

3. Methods of Preparation

This preparation of 2-methoxy-1,8-diphenyl-1a,2,7,7a-tetrahydro-1,2,7-metheno-1*H*-cyclopropa(b)naphthalene (I) is based on the published procedure.[2] (I) has also been prepared from 4-methoxy-1,2-diphenyl-2a,8b-dihydrocyclobuta(a)naphthalene.[3]

4. Merits of the Preparation

The present procedure, which uses commercially available starting materials, has been applied to a wide range of substituted naphthalenes[4] and to several heterocyclic diarylacetylenes.[5,6]

References

1. W. H. F. Sasse, *Aust. J. Chem.*, **22**, 1257 (1969).
2. W. H. F. Sasse, P. J. Collins. D. B. Roberts, and G. Sugowdz, *Aust. J. Chem.*, **24**, 2339 (1971).
3. P. J. Collins and W. H. F. Sasse, *Aust. J. Chem.*, **24**, 2325 (1971).
4. W. H. F. Sasse, P. J. Collins, D. B. Roberts, and G. Sugowdz, *Aust. J. Chem.*, **24**, 2151 (1971).
5. T. Teitei, P. J. Collins, and W. H. F. Sasse, *Aust. J. Chem.*, **25**, 171 (1972).
6. T. Teitei, D. Wells, and W. H. F. Sasse, *Aust. J. Chem.*, **26**, 2129 (1973).

METHYLENECYCLOPENTANE

$$BH_3 + \text{cyclopentene} \longrightarrow (\text{C}_5\text{H}_9)_3B \xrightarrow[H_3O^+]{C_6H_5\overset{O}{\overset{\|}{C}}CH{=}CH_2} \text{C}_5\text{H}_9CH_2CH_2\overset{O}{\overset{\|}{C}}-C_6H_5$$

$$\text{C}_5\text{H}_9CH_2CH_2\overset{O}{\overset{\|}{C}}-C_6H_5 \xrightarrow{h\nu} \text{C}_5\text{H}_8{=}CH_2 + C_6H_5-\overset{O}{\overset{\|}{C}}CH_3$$

Submitted by D. C. NECKERS,* R. M. KELLOGG,† W. L. PRINS,† and B. SCHOUSTRA*
Checked by T. D. ROBERTS and R. MILLER‡

1. Procedure

Diborane (1 M, 100 ml) (Note 1) is added to predried cyclopentene (20.4 g, 0.3 mole) at 0° (Note 2).[1,2] The solution is warmed to 25° for 1 hour, after which water (4.5 g, 0.25 mole) is added with vigorous stirring. Following this hydrolysis phenyl vinyl ketone (23.7 g, 0.18 mole) in 100 ml of THF is added (Notes 3 and 4). After 1 hour at 25° the THF is removed and the reaction mixture held at 60°/15 torr for 1 hour to remove all volatile components, leaving crude 3-cyclopentyl-1-phenylpropanone-1, which is used without further purification (Note 5).

For the irradiation a 500-ml photoreaction vessel (available from Ace Glass) is used. A distillation head is attached to the irradiation vessel. A condenser is connected to a collection flask and to a trap cooled in liquid nitrogen. A vacuum pump and a manostat complete the apparatus. Dry nitrogen is passed through the system during irradiation.

A suspension of 58 g of the crude ketone (Note 6), above, is placed in sufficient (*ca.* 500 ml) tetramethylene glycol dimethyl ether to fill the irradiation vessel. The mixture is stirred vigorously and irradiated for 25 hours at 40° under a controlled vacuum of 3 torr. Methylenecyclopentane (16 g, 68%) is collected in the liquid nitrogen trap (Notes 7 and 8).

* Bowling Green University, Bowling Green, Ohio 43403.
† Department of Organic Chemistry, The University, Zernikelaan, Groningen, The Netherlands.
‡ University of Arkansas, Fayetteville, Arkansas 72701.

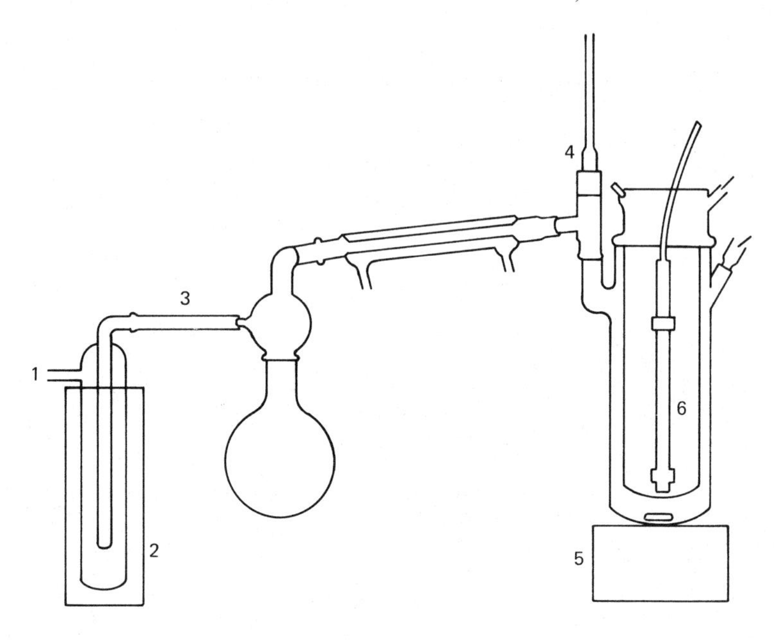

1. To vacuum pump. 2. Liquid nitrogen Dewar. 3. Teflon connection. 4. Thermometer, 5. Magnetic stirrer, 6. Hanovia 450-watt mercury arc lamp in Ace Glass irradiation vessel (Pyrex).

2. Notes

1. Diborane, approximately 1 M in THF, is purchased from either Alfa Inorganics or Ventron Corporation. It can be used directly.

2. All operations were carried out in a 500-ml three-necked flask provided with a stirrer, dropping funnel, and condenser. The entire apparatus was kept under a positive pressure of nitrogen during the operations. The checkers found that the reaction took 12 hours with vigorous stirring.

3. Phenyl vinyl ketone is prepared from β-diethylaminopropiophenone.[3] Because phenyl vinyl ketone polymerizes easily, it is prepared fresh before each use. A steam distillation apparatus is charged with 24.1 g (0.1 mole) of β-diethylaminopropiophenone hydrochloride and steam distillation is continued until the distillate becomes clear. The distillate is extracted with 3×50 ml of ether and the extracts are dried over $MgSO_4$. Removal of the solvent leaves 12.0 g (0.091 mole, 91%) of phenyl vinyl ketone which is used without further purification.

4. The reaction of trialkyl borane with phenyl vinyl ketone is quite vigorous and cooling is necessary to maintain the temperature at 25°.

5. 3-Cyclopentyl-1-phenylpropanone-1: nmr (CCl_4) δ 0.9–2.0 (complex m, 11) 2.95 (t, J = 7.0 Hz, 3) 7.3–8.0 (complex m, 5). 2,4-dinitrophenylhydrazone, m.p. 157–159°.

6. Most higher molecular weight ketones are not soluble in tetramethylene glycol dimethyl ether. (The checkers used triglyme.) The progress of the reaction can be monitored by watching the suspension disappear (with the lamp off to avoid injury to the eyes).

7. Nitrogen flow removes oxygen, a quencher of triplet state processes. Vacuum is used to remove the product as it forms and to prevent oxetane formation between the various ketones and the olefinic products. The checkers used three Dry Ice-acetone traps in succession instead of the liquid nitrogen trap. The product collected in the first two traps.

8. The methylenecyclopentane obtained by photochemical synthesis was free of contaminants as determined by gas chromatography and nmr spectra. Spectral data are ir (neat) 3080, 2920, 1045, and 885 cm^{-1}; nmr (CCl_4) δ 1.4–1.9 (broad s, 4, —CH_2CH_2—), δ 2.0–2.3 (complex m, 4, =C—CH_2—), and δ 4.71 (broad 2, vinyl H).

3. Discussion

The synthetic utility of the Norrish Type II photoelimination reaction has been discussed in detail by Neckers, Kellogg, and coworkers.[4] The method is limited under its present application to olefins of sufficient volatility to distill at 3 torr/40°.

The major advantage of the Norrish Type II reaction is that stereochemically pure olefins are produced. In that regard multiple elimination sequences have been reported with long-chain arylalkyl diones which produce dienes of specific orientation. Butadiene, 1,4-pentadiene, and 1,5-hexadiene have all been prepared in this way.

References

1. H. C. Brown, *Hydroboration*, Benjamin, New York, 1962, p. 98; A. Zuzaki, A. A. H. Matsumato, M. Itoh, H. C. Brown, M. M. Rogie, and M. W. Rathke, *J. Am. Chem. Soc.*, **89**, 5708 (1967).
2. G. Zweifel and J. Plamondon, *J. Org. Chem.*, **35**, 898 (1970).
3. C. E. Maxwell, *Org. Synth.*, Collect. Vol. 3, 305 (1955).
4. D. C. Neckers, R. M. Kellogg, W. L. Prins, and B. Schoustra, *J. Org. Chem.*, **36**, 1838 (1971).

4-METHYLTRICYCLO[5.3.1.0^{5,10}]UNDEC-2-ENE-6,9-DIONE

Submitted by JOHN R. SCHEFFER and RUDOLF E. GAYLER*
Checked by J. CORNELISSE and J. N. M. BATIST†

1. Procedure

A. 5α,8α-Dimethyl-4aβ,5,8,8aβ-tetrahydro-1,4-naphthoquinone. A mixture of *p*-benzoquinone (3.1 g, 0.029 mole) (Note 1) and *trans,trans*-2,4-hexadiene (5.6 g, 0.068 mole) (Note 2) is heated at 60° for 30 minutes, followed by stirring at room temperature for 3 hours. Removal of excess diene by vacuum rotary evaporation, followed by recrystallization of the resulting solid from hexane, gives 3.1 g (56%) of the desired Diels-Alder adduct, m.p. 57.0–57.5° (light yellow) (lit.[1] 58–59.5°). Concentration and crystallization of the mother liquor produces a second crop of 1.7 g (31%), m.p. 51–53°.

B. 4-Methyltricyclo[5.3.1.0^{5,10}]undec-2-ene-6,9-dione. The Diels-Alder adduct prepared above (1.0 g, 5.3 mmoles) in 400 ml of benzene (Note 3) is placed in a cylindrical Pyrex immersion-well flask fitted with a magnetic stirring bar, a fritted inert gas inlet disc, and a stopcock-regulated sampling device. After thorough deoxygenation with high purity nitrogen (2 hours) the solution is irradiated *externally* from a distance of 1 ft by using a 275 W Westinghouse uv sun lamp (Note 4) with a 16.5 × 16.5 cm sheet of Corning No. 7380 glass (transmitting $\lambda > 340$ nm) interposed between the lamp and the reaction vessel (Note 5). The reaction may be conveniently followed by monitoring the decrease in the uv absorption of the starting material at 371 nm (Note 6). The reaction is complete when no further change is observed in the uv spectrum in this region (120 hours). The crude photolysate is concentrated *in vacuo* and the resulting dark brown oil subjected to column chomatography on neutral Woelm alumina with chloroform as the eluting solvent (Note 7). The homogeneous oil thus obtained is dissolved

* University of British Columbia, Vancouver 8, Canada.
† Gorlaeus Laboratoria, Rijksuniversiteit, Leiden, The Netherlands.

in hexane, whereupon crystals (275 mg, 28%) of 4-methyltricyclo[5.3.1.$0^{5,10}$]-undec-2-ene-6,9-dione, m.p. 75.5–78.5°, are deposited. Two further recrystallizations give analytically pure material, m.p. 80.0–80.5°.

Similar results are obtained by using hexane or *tert*-butyl alcohol as the photolysis solvent. In the latter solvent reaction is complete within 24 hours.

2. Notes

1. Eastman Organic Chemicals Practical Grade benzoquinone was used.
2. Obtained from Aldrich Chemical Co.
3. The benzene was Fisher Certified A.C.S. Grade distilled before use.
4. This lamp, commonly used for suntanning, is readily available in most drugstores in package form which includes a convenient lamp stand. The checkers used a Rayonet type RS reactor fitted with 3500 Å lamps.
5. Obtained from Corning Glass Works, Corning, New York, 14830. C.S. No. 0-52.
6. The concentration of the photolysis solution is such that aliquots can be analyzed directly in this wavelength region by using 1-cm cells.
7. Column size, 1.4 × 15 cm.

3. Discussion

The photochemical reaction described is likely to be general for butadiene benzoquinone Diels-Alder adducts bearing α-substituents at C_5 and C_8 with at least one γ-hydrogen atom; that is, —CHR_1R_2, —CH_2R, or —CH_3.[2] Unsymmetrical adducts with only one C_5 or C_8 substituent (e.g., the piperylenebenzoquinone Diels-Alder adduct) also undergo this reaction. The mechanism has been shown to consist of three steps: (a) γ-hydrogen abstraction from a methyl group, (b) bonding between the resulting methylene radical and C_3, and (c) ketonization of the enol thus formed.

The procedure is by far the simplest entry into the tricyclo[5.3.1.$0^{5,10}$]-undecane ring system found in patchouli alcohol[3] and seychellene[4] and their derivatives.

References

1. H. V. Euler; H. Hasselquist, and A. Glaser, *Arkiv Kemi*, **3**, 49 (1952).
2. J. R. Scheffer, K. S. Bhandari, R. E. Gayler, and R. H. Wiekenkamp, *J. Am. Chem. Soc.*, **94**, 285 (1972).
3. G. Büchi, W. D. MacLeod, Jr., and J. Padilla O., *ibid.*, **86**, 4438 (1964).
4. G. Wolff and G. Ourisson, *Tetrahedron*, **25**, 4903 (1969).

5-NORBORNENE-2-*endo*-CARBOXYLIC ACID AND 5-NORBORNENE-2-*endo*-METHANOL

CO_2H $\xrightarrow[NaHCO_3]{I_2}$ I ... O ... O + CO_2H

hv
CH_3OH
Zn

CO_2H $\xrightarrow[(2)\ LiAlH_4]{(1)\ CH_2N_2}$ CH_2OH

Submitted by TAPPEY H. JONES and PAUL J. KROPP*
Checked by J. CORNELISSE and Mrs. G. M. GORTER-LAROY†

1. Procedure

In a quartz tube (45 cm long and 4.8 cm in diameter) are placed a 300-ml methanolic solution containing hexahydro-6-iodo-3,5-methano-2*H*-cyclopenta(b)furan-2-one (20.0 g, 76 mmoles), m.p. 57–58°, prepared as previously detailed,[1] and fine mossy zinc (5 g) (Note 1). The mixture is stirred magnetically, purged with nitrogen, and irradiated for 8 hours in a Rayonet Photochemical Reactor containing 16 G8T5 low-pressure mercury lamps (Note 2). The solution is then filtered and the solvent is removed under reduced pressure. After the addition of 150 ml of diethyl ether and treatment with 50 ml of saturated sodium bicarbonate solution the mixture is carefully acidified with 10% hydrochloric acid; the aqueous layer is then separated and washed twice with 50 ml of diethyl ether. The combined ether extracts are dried over anhydrous sodium sulfate and concentrated to a total volume of about 50 ml (Note 3a).

To this solution is added an ethereal solution of diazomethane[2] (from 11.0 g of *N*-nitroso-*N*-methylurea) (Note 4) until the yellow color persists.

* Department of Chemistry, University of North Carolina, Chapel Hill, North Carolina 27514.
† Gorlaeus Laboratoria, Rijksuniversiteit, Leiden, The Netherlands.

After evaporation of diazomethane the solution is reduced in volume to about 50 ml and dried over anhydrous sodium sulfate (Note 3b).

The solution is then added dropwise to a suspension of lithium aluminum hydride (5 g) in 150 ml of anhydrous diethyl ether which has been purged with nitrogen. The resulting mixture is refluxed for 30 minutes and then allowed to cool. Excess lithium aluminum hydride is quenched by dropwise addition of ethyl acetate and the resulting mixture is treated with water and dilute sodium hydroxide solution.[3] The precipitated salts are washed several times with ether; the ether solution is dried over anhydrous sodium sulfate; and the solvent is removed under reduced pressure. Distillation affords 7.7–7.8 g (82–83%) of a colorless oil, b.p. 95–97° (15 mm), which is 97% pure by gas-chromatographic analysis (Note 5).

2. Notes

1. Zinc metal is used as an iodine scavenger. Similar results are obtained in the absence of zinc except that the reaction is slowed down by the build-up of iodine, which competes for the light.
2. Manufactured by the Southern New England Ultraviolet Co., 954 Newfield Street, Middletown, Connecticut, Cat. No. RPR-100.
3. At this point an intermediate product may be obtained by vacuum distillation under the following conditions: (a) 5-norbornene-2-*endo*-carboxylic acid, b.p. 129–132° (18 mm); (b) methyl 5-norbornene-2-*endo*-carboxylate, b.p. 68–71° (11 mm).
4. This step should be performed in a well-ventilated hood.
5. A column, 10-ft × 0.25 in., containing 20% Carbowax 20M on 60/80 mesh Chromosorb W is used.

3. Methods of Preparation

β-Iodolactones are also cleaved by treatment with zinc and acetic acid.[4]

4. Merits of the Preparation

Iodolactonization, followed by regeneration of the acid, is a useful technique for the resolution of epimeric mixtures of β,γ-unsaturated carboxylic acids like those obtained from the Diels-Alder reaction.[1,4] This procedure provides a useful alternative to the traditional method for regenerating the unsaturated acid, which requires a tedious isolation of the acid from acetic acid.[4] β-Iodoethers can also be cleaved by this photochemical method.[5] Moreover, it represents a general procedure for the photoconversion of alkyl iodides to cationic intermediates, using zinc as an iodine scavenger.[5]

References

1. C. D. Ver Nooy and C. S. Rondestvedt, Jr., *J. Am. Chem. Soc.*, **77**, 3583 (1955).
2. F. Arndt, *Org. Synth. Collect. Vol.* **2**, 1965 (1943).
3. L. F. Fieser and M. Fieser, *Reagents for Organic Synthesis*, Vol. 1, Wiley, New York, 1967, p. 584.
4. C. S. Rondestvedt, Jr., and C. D. Ver Nooy, *J. Am. Chem. Soc.*, **77**, 4878 (1955).
5. P. J. Kropp, T. H. Jones, and G. S. Poindexter, *ibid.*, **95**, 5420 (1973).

5-(2-OXYPROPYL)-6-AZAURACIL

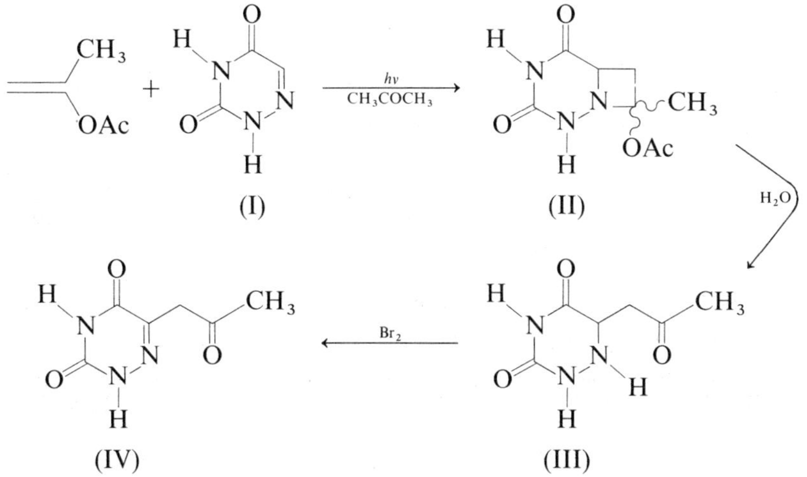

Submitted by J. S. SWENTON and R. J. BALCHUNIS*
Checked by T. D. ROBERTS and C. T. ROBERTS†

1. Procedure

6-Azauracil (I), 2.0 g (0.018 mole) (Note 1), is dissolved in 200 ml of warm 70% acetone-water and 6 ml (0.06 mole) of isopropenyl acetate is added. The solution is magnetically stirred and irradiated for 2.5 hours with Corex-filtered light from a 450-watt Hanovia medium-pressure mercury lamp (Note 2) while nitrogen is slowly bubbled through the solution. At the end of the irradiation the clear solution is concentrated *in vacuo* on the rotary evaporator at about 70° until a white crystalline solid separates from the solution (Note 3). The evaporation is then interrupted and the mixture chilled overnight. The resulting pure white solid is collected by filtration and dried *in vacuo* at room temperature to yield pure (III), 2.3 g (77%) m.p. 189–191°. No further purification is required for the next step (Note 4).

The dihydro-6-azauracil (III) is dissolved in 100 ml of water at room temperature. The solution is vigorously stirred while bromine is added dropwise until a light yellow color persists. After 10–15 minutes the reaction mixture turns colorless again and the solution is concentrated on the rotary evaporator to about 25 ml. After refrigeration white needles form,

* Ohio State University, Columbus, Ohio 43210.
† University of Arkansas, Fayetteville, Arkansas 72701.

which are collected and dried *in vacuo* to provide 5-(2-oxypropyl)6-azauracil (IV) 1.52 g (76%), m.p. 173–176°C (Note 5).

2. Notes

1. The 6-azauracil is commercially available from Aldrich Chemical, Milwaukee, Wisconsin.
2. The apparatus is essentially as described in *Organic Photochemical Syntheses*, Vol. I, p. 14.
3. The initially formed bicyclic azetidine (II) is readily hydrolyzed in the hot aqueous solution to (III). In the case of 1,3-dimethyl-6-azauracil photoadditions the epimeric bicyclic azetidines have been isolated and characterized.[1]
4. This material can be recrystallized from water-ethanol to give a purified product of m.p. 191–192.5°C.
5. In some runs this material develops a slight discoloration on standing. The discoloration can be removed by recrystallization from ether-ethanol to give material m.p. 174–176°C.

3. Methods of Preparation

Published methods for preparing 5-aryl or 5-alkyl-6-azauracil derivatives involve ring-closure reactions of thiosemicarbazones of appropriate α-keto acids.[2,3] To our knowledge this is the only method available for directly functionalizing the parent 6-azauracil.

4. Merits of the Preparation

This reaction serves as an exceptionally mild, high-yield method of preparing 5-functionalized 6-azauracils. The submitters have found that the reaction works well also with cyclohexenyl acetate and presumably most simple enol acetates should undergo smooth addition. The route outlined here suggests that similar reactions for other conjugated imine systems would also be of synthetic utility.

References

1. J. S. Swenton and J. A. Hyatt, *J. Am. Chem. Soc.*, **96**, 4879 (1974).
2. J. Grut, *Adv. Heterocycl. Chem.*, **1**, 204 (1963).
3. M. Bobek, J. Farkas, and J. Grut, *Coll. Czech. Chem. Comm.*, **32**, 1295 (1967).

2-PHENYL-4-CYANO-5,5-DIMETHYL-Δ[1]-PYRROLINE

$$CO_2 + \text{3,3-dimethyl-2-phenylazirine} \xrightarrow{h\nu} \text{2,2-dimethyl-4-phenyl-}\Delta^3\text{-oxazolin-5-one}$$

$$CH_2{=}CHCN + \text{2,2-dimethyl-4-phenyl-}\Delta^3\text{-oxazolin-5-one} \xrightarrow{h\nu} \text{2-phenyl-4-cyano-5,5-dimethyl-}\Delta^1\text{-pyrroline} + CO_2$$

Submitted by ALBERT PADWA and S. I. WETMORE, JR.*
Checked by T. WOOLDRIDGE and T. D. ROBERTS†

1. Procedure

A. 2,2-Dimethyl-4-Phenyl-Δ³-Oxazolin-5-one. The apparatus consists of a water-cooled immersion well containing a Corex filter and a 450-watt Hanovia medium-pressure mercury arc lamp. The flask is fitted with a side arm to a condenser, a gas inlet tube, and an outlet fitted with polyethylene tubing to allow for removal of aliquots. The solution is stirred magnetically.

The photolysis is carried out by dissolving 3,3-dimethyl-2-phenylazirine (0.5 g) (Note 1) in 250 ml of freshly distilled benzene which is saturated with carbon dioxide. The solution is photolyzed as carbon dioxide is continuously bubbled through the photolysis well (Note 2). The irradiation is followed by monitoring the disappearance of the nmr absorption bands of the azirine (Note 3). After 1.5 hours (Note 4) nmr analysis indicates the quantitative conversion of starting material to a new product. The solvent is evaporated on a rotary evaporator under reduced pressure at *ca.* 30°. The residue is distilled at 100–105° (0.02 mm) to yield 0.61 g (94%) of a clear oil which solidifies in the receiver. The white solid obtained is recrystallized twice from hexane to yield an analytical sample of 2,2-dimethyl-4-phenyl-Δ³-oxazolin-5-one, m.p. 36–37°; nmr ($CDCl_3$, 100 MHz) τ 8.34 (6*H*, s), 2.3–2.5 (3*H*, m), and 1.4–1.6 (2*H*, m).

B. 2-Phenyl-4-Cyano-5,5-Dimethyl-Δ¹-Pyrroline. The source of uv (2537 Å) radiation is a Rayonet reactor (Note 5). The apparatus consists of a Griffin-Worden quartz pressure vessel (Note 6), 26.0 cm long and 2.1 cm

* State University of New York, Buffalo, New York 12214.
† University of Arkansas, Fayetteville, Arkansas 72701.

O.D., closed at one end and fitted at the open end with an end cap. A solution of 2,2-dimethyl-4-phenyl-Δ^3-oxazolin-5-one (0.3 g) and acrylonitrile (2.0 ml) in 60 ml of pentane is irradiated at room temperature for 5 hours (Note 8). Removal of the solvent under reduced pressure affords an oily residue that is recrystallized from hexane:ether to give 0.21 g (66%) of 2-phenyl-4-cyano-5,5-dimethyl-Δ^1-pyrroline as a crystalline solid, m.p. 65–66° (Note 7); nmr ($CDCl_3$, 100 MHz) τ 8.56 (3*H*, s), 8.50 (3*H*, s), 7.09 (1*H*, *dd*, $J = 10$, and 8.0 Hz), 6.76 (1*H*, *dd*, $J = 17$, and 8.0 Hz), 6.58 (1*H*, *dd*, $J = 17$, and 10.0 Hz), and 2.2–2.8 (5*H*, m).

2. Notes

1. The 3,3-dimethyl-2-phenylazirine was prepared as described by Leonard and Zwanenburg,[1] b.p. 93–94° (15 mm); nmr (CCl_4) τ 8.65 (6*H*, s), 2.2–2.6 (5*H*, m).
2. Solid carbon dioxide was placed in an Erlenmeyer flask which, in turn, was fitted with a stopper connected to a polyethylene tube by means of a glass rod. The tubing was connected to the gas inlet tube of the photolysis apparatus and a positive pressure of carbon dioxide was maintained during the course of the irradiation.
3. The irradiation was stopped after the methyl singlet of the azirine ring (τ 8.65) had disappeared.
4. Prolonged irradiation resulted in the formation of a mixture of secondary photoproducts.
5. Supplied by Southern New England Ultraviolet Co., Middletown, Connecticut.
6. The Griffin-Worden quartz pressure vessel was obtained from Kontes Glass Co., Vineland, New Jersey, Cat. No. K-767150.
7. 2-Phenyl-4-cyano-5,5-dimethyl-Δ^1-pyrroline can also be prepared by irradiating a mixture of 3,3-dimethyl-2-phenylazirine with acrylonitrile, using a 450-watt mercury Hanovia lamp equipped with a Pyrex filter sleeve.[2]
8. The checkers observed that a white solid precipitated slowly during photolysis and gradually coated the inside of the reaction vessel. It was removed by filtration and discarded.

3. Methods of Preparation

Arylazirines undergo photocycloaddition with a variety of electron-deficient olefins to give Δ^1-pyrroline derivatives.[2,3] The formation of the adducts can be interpreted as proceeding by opening the azirine ring to form a nitrile ylide which is subsequently trapped by a suitable dipolarophile. The photocycloaddition reaction is not restricted to just electron-deficient double bonds but also occurs with a wide variety of hetero-multiple bonds[4,5] and provides a convenient route for the synthesis of five-membered heterocyclic rings.

4. Merits of the Preparation

Thermal 1,3-dipolar cycloadditions of Δ^2-oxazolin-5-ones with electron-deficient olefins to give Δ^1-pyrrolines are known.[6] The reaction of benzimidoyl chlorides with trimethylamine in the presence of a dipolarophile has also been shown to produce Δ^1-pyrroline derivatives.[7] The present procedure has the advantage of using readily available starting material with a wide assortment of substituent groups. The reaction and purification are relatively simple procedures to carry out. The general method has been used successfully in the submitter's laboratory for the preparation of a wide assortment of variously substituted five-membered heterocyclic rings. The nonphotochemical routes require elevated temperatures[6] or basic reaction conditions[7] and work only with a selected number of substituent groups.

References

1. N. Leonard and B. Zwanenburg, *J. Am. Chem. Soc.*, **89**, 4456 (1967).
2. A. Padwa, M. Dharan, J. Smolanoff, and S. I. Wetmore, Jr., *J. Am. Chem. Soc.*, **95**, 1945 (1973).
3. A. Padwa, M. Dharan, J. Smolanoff, and S. I. Wetmore, Jr., *J. Am. Chem. Soc.*, **95**, 1954 (1973).
4. A. Padwa, J. Smolanoff, and S. I. Wetmore, Jr., *J. Org. Chem.*, **38**, 1333 (1973).
5. B. Jackson, N. Gakis, M. Marky, H. J. Hansen, W. von Philipsborn, and H. Schmid, *Helv. Chim. Acta*, **55**, 916 (1972).
6. H. Gotthardt, R. Huisgen, and H. O. Bayer, *J. Am. Chem. Soc.*, **92**, 4340 (1970).
7. R. Huisgen, H. Stangl, H. J. Sturm, and H. Wagenhofer, *Chem. Ber.*, **105**, 1258 (1972).

1-PHENYLSPIRO[2.6]NONA-4,6,8-TRIENE

$$\text{Tropone} \xrightarrow[\text{(3) NaH}]{\text{(1) SOCl}_2\text{; (2) ArSO}_2\text{NHNH}_2} \text{Tropone}{=}\text{N}\bar{\text{N}}\text{SO}_2\text{Ar}\ \text{Na}^+$$

$$\text{Tropone}{=}\text{N}\bar{\text{N}}\text{SO}_2\text{Ar}\ \text{Na}^+ + \text{PhHC}{=}\text{CH}_2 \xrightarrow[\text{THF}]{h\nu} \text{1-phenylspiro[2.6]nona-4,6,8-triene (Ph)}$$

Submitted by RUSSELL A. LaBAR*
Checked by J. CORNELISSE, G. M. GORTER-LAROY and J. N. M. BATIST†

1. Procedure

Tropone may be purchased from Koch-Light Laboratories, Ltd. England, and should be distilled before use (Note 1). Tropone tosylhydrazone[1] is prepared by adding tropone dropwise to an excess of thionyl chloride at 0°, followed by heating at reflux for 5 minutes. Removal of the excess thionyl chloride at reduced pressure produces a yellow crystalline dichloride which is used without further purification. An equal molar amount of *p*-toluenesulfonylhydrazine in a small amount of abs. ethanol is added to the dichloride in abs. ethanol and stirred at room temperature for 45 minutes. The solution is cooled to 0° and the yellow precipitate (tropone *p*-toluenesulfonylhydrazone hydrochloride) is vacuum filtered. The hydrochloride salt is added slowly with vigorous stirring to a mixture of 500 ml saturated sodium bicarbonate and 300 ml methylene chloride. The dark methylene chloride layer is removed and the aqueous layer is washed twice with 150 ml portions of methylene chloride. The combined organic layers are dried and concentrated on a rotary evaporator. The crude residue is recrystallized from benzene/*n*-pentane to yield the tosylhydrazone, m.p. 142–143°.

The sodium salt is prepared by slowly adding (in a dry box) an equal molar amount of sodium hydride (57% dispersion in mineral oil) to the tosylhydrazone in dry THF. After complete addition the mixture is stirred for 30 minutes more and then vacuum filtered. The resulting solid (sodium salt) is washed with dry THF and dried. The salt may be stored indefinitely at room temperature in a dry box.

Into a cylindrical, water-cooled Pyrex photolysis apparatus (Fig. 1) is placed 100 ml of distilled dry THF and 100 ml of freshly distilled styrene.

* Department of Chemistry, University of Florida, Gainesville, Florida 32611.
† Gorlaeus Laboratoria, Rijksuniversiteit, Leiden, The Netherlands.

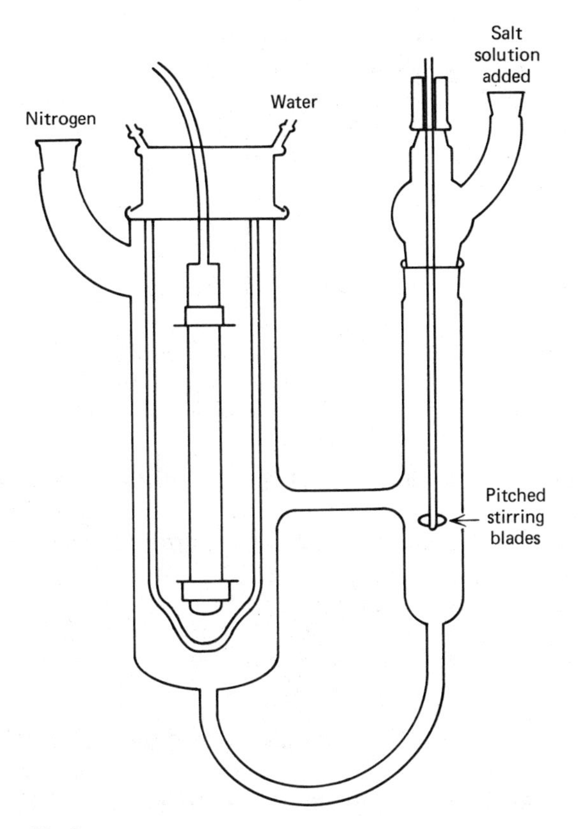

Fig. 1.

The temperature of the reaction solution is approximately 30° and a positive pressure of nitrogen is maintained throughout the photolysis. To the rapidly stirred, irradiated (Note 2) mixture a solution of 0.5–0.6 g (*ca.* 2.0 mole) of the sodium salt of tropone tosylhydrazone in 15 ml of dry DMF is added dropwise over a period of 2–3 hours. If, during addition of the salt, the reaction solution in the sidearm with stirrer turns brown, the addition is discontinued and the solution is photolyzed until it returns to a yellow color. After complete addition of the salt the reaction mixture is irradiated for an additional hour.

The resulting solution is poured into 500 ml of water and extracted three times with 150-ml portions of pentane. The combined pentane extract is dried (Na_2SO_4) and concentrated on a rotary evaporator, and the excess styrene and DMF are removed by vacuum distillation (Note 3). The brown oil is first percolated through a short basic alumina column with 20:80

ether/pentane to remove traces of polystyrene and then carefully chromatographed on a basic alumina (activity I) column measuring 20 × 1.6 cm. Elution with 5% ether/pentance gives a slow-moving pale yellow band (Note 4). Removal of the solvent yields 1-phenylspiro[2.6]nona-4,6,8-triene (Note 5) as a clear pale yellow oil in 45–52% yield.

2. Notes

1. The checkers prepared tropone according to P. Radlich, *J. Org. Chem.*, **29**, 960 (1964).
2. A Hanovia 550-watt mercury arc lamp is used for irradiation.
3. The temperature during irradiation and workup should never exceed 50° because of the facile thermal rearrangement of the product to 8-phenylbicyclo[5.2.0]nona-1,3,5-triene.[2]
4. A trailing brown band caused by the formation of dimeric heptafulvalene may be observed. Generally all the eluent preceding this band is collected. Traces of styrene may be present after removal of the solvent and can be removed at reduced pressure.
5. Spectral data are ir (neat) 3080, 3055, 3010, 1601, 1495, 1451, 909, 776, 732, and 697 cm^{-1}; nmr ($CDCl_3$) δ 1.21 (d,2), 2.10 (t,1), 5.00 (d,1), 5.35 (d, 1), 5.9 to 6.5 (m,4), and 7.1 (m,5).

3. Methods of Preparations

The reported synthesis of 1-phenylspiro[2.6]nona-4,6,8-triene utilizes this general procedure.[2,3]

4. Merits of Preparation

This procedure provides a general method of synthesis of the spiro[2.6]-nona-4,6,8-triene system. Other alkenes which have been successfully used include dimethyl fumarate and maleate,[4] fumaronitrile and maleinitrile.[1]

References

1. W. M. Jones and C. L. Ennis, *J. Am. Chem. Soc.*, **91**, 6391 (1969).
2. E. E. Waali and W. M. Jones, *J. Am. Chem. Soc.*, **95**, 8114 (1973).
3. L. W. Christensen, E. E. Waali, and W. M. Jones, *J. Am. Chem. Soc.*, **94**, 2118 (1972); K. Untch, private communication.
4. W. M. Jones, B. N. Hamon, R. C. Joines, and C. L. Ennis, *Tetrahedron Lett.*, 3909 (1969).

4-PHENYLPHENANTHRENE AND 3-IODO-4-PHENYLPHENANTHRENE

$+I_2$ $\xrightarrow{h\nu}$

Submitted by A. H. A. TINNEMANS and W. H. LAARHOVEN*
Checked by Th. J. H. M. CUPPEN and R. J. F. M. van ARENDONK*

1. Procedure

A. 4-Phenylphenanthrene. A solution of 1-(2-naphthyl)-4-phenylbut-1-en-3-yne (2.5 g, 0.01 mole) (Note 1) in 1 liter of methanol (Note 2) is placed in a quartz tube (5 cm in diameter) and irradiated for 50 hours (Note 3) in a Rayonet RPR.-100 reactor fitted with 300-nm lamps. The reaction mixture is transferred to a 2-liter, round-bottomed flask and evaporated to dryness on a rotary evaporator. The pale yellow residue is dissolved in 25 ml of carbon tetrachloride and the solution poured onto a column of silica gel (Merck Kieselgel 60) 3 cm in diameter and 10 cm long. The column is eluted with three 100-ml portions of hexane-toluene (4:1). The elution is stopped before a yellow band, which is developed, reaches the end of the column. The combined eluates are evaporated to dryness leaving 1.40 g of crude product. Crystallization from methanol yields 1.25 g (50%) of pure 4-phenylphenanthrene, m.p. 82–84°C (Note 4).

B. 3-Iodo-4-Phenylphenanthrene. To a solution of 1-(2-naphthyl)-4-phenylbut-1-en-3-yne (2.5 g, 0.01 mole) in 1 liter of benzene is added

* Department of Organic Chemistry, University of Nijmegen, Toernooiveld, Nijmegen, The Netherlands.

iodine (2.5 g, 0.01 mole). The solution is placed in a Pyrex tube and irradiated for 20 hours at 360 nm in a Rayonet RPR.-100 reactor (Notes 3 and 5). After evaporation of the solvent the residue is chromatographed over a silica gel column 3 cm in diameter and 15 cm long by using two 250-ml portions of hexane-toluene (65:35) as eluent. The combined eluates are evaporated to dryness. The yellow residue is dissolved in carbon tetrachloride and again chromatographed over a column of alumina (Merck, aluminium oxide, neutral, activity 1) 3 cm in diameter and 20 cm long. Elution with 100 ml of hexane-toluene (9:1) gives 4-phenylphenanthrene in about 5% yield. A second elution with 500 ml of hexane-toluene (85:15) gives 2.6 g (69%) of the crude product. Crystallization from methanol yields 2.2 g (58%) of 3-iodo-4-phenylphenanthrene, m.p. 157–160° (Note 6).

2. Notes

1. 1-(2-naphthyl)-4-phenylbut-1-en-3-yne can be prepared by a Wittig reaction from phenylpropargylaldehyde[2] and the triphenylphosphonium salt of 2-bromomethylnaphthalene. Each of the isomers (*trans*, m.p. 131–132°; *cis*, m.p. 96–97°) and a mixture of them can be used for the synthesis of 4-phenylphenanthrene.
2. With other suitable solvents, such as hexane, cyclohexane, and benzene, the required irradiation time is slightly different (see Note 3).
3. The reaction can be followed by uv spectroscopy by using the decrease of the absorption of the starting material at 340–350 nm as a measure of the progress of the reaction (*trans*, λ_{max} 345 nm, $\varepsilon = 35{,}800$; *cis*, λ_{max} 340 nm, $\varepsilon = 18{,}500$; 4-phenylphenanthrene, λ_{max} 349 nm, $\varepsilon = 380$).
4. Reported melting point: 80.5–81.5°.[1]
5. 3-Iodo-4-phenylphenanthrene; uv: λ_{max} 345 nm, $\varepsilon = 340$; λ_{max} 354 nm, $\varepsilon = 530$.
6. Repeated crystallization gives m.p. 159–160°. After sublimation a melting point of 161–162° was obtained.

3. Methods of Preparation

These preparations are based on the procedure published by the submitters.[3] 4-Phenylphenanthrene has already been prepared in 18% yield by irradiation of 1-(2-naphthyl)-4-phenylbuta-1,3-diene[4] and in 5% overall yield by refluxing a benzene solution of phenylpropionylchloride and β-2-naphthylpropiolic acid and decarboxylation of the isolated 4-phenylphenanthrene-2,3-dicarboxylic anhydride.[1]

The photoisomerization of 1,4-diarylbutenynes can be used for the preparation of several other phenylaromatics and their iodo derivatives which are otherwise difficult to prepare; for example, 1-phenylphenanthrene from 1-(α-naphthyl)-4-phenylbut-1-en-3-yne; 1-phenylbenzo[c]-phenanthrene from 1-(3-phenanthryl)-4-phenylbut-1-en-3-yne; 4,5-diphenylphenanthrene from 1-(1-phenyl-7-naphthyl)-4-phenylbut-1-en-3-yne.[5]

References

1. A. D. Campbell, *J. Chem. Soc.*, 3659 (1954).
2. *Org. Synth. Collect. Vol.* **3**, 731 (1965).
3. A. H. A. Tinnemans and W. H. Laarhoven, *Tetrahedron Lett.*, 817 (1973).
4. R. J. Hayward and C. C. Leznoff, *Tetrahedron*, **27**, 2085 (1971).
5. A. H. A. Tinnemans and W. H. Laarhoven, *J. Am. Chem. Soc.*, **96**, 4617 (1974).

2,2,6,6-TETRAMETHYLCAPROLACTAM

$$\xrightarrow{h\nu}$$

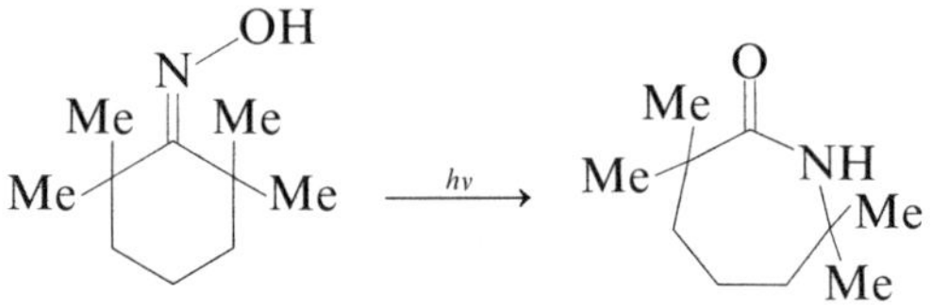

Submitted by GEORGE JUST and MICHAEL CUNNINGHAM*
Checked by T. D. ROBERTS†

1. Procedure

A solution of the oxime (m.p. 151.5°C) of 2,2,6,6-tetramethylcyclohexanone[1,2] (1 mmole) in methanol (100 ml) is placed in a quartz cell and nitrogen is bubbled through the solution for 20 minutes (the nitrogen atmosphere is maintained by a mercury seal). The solution is photolyzed in a Rayonet Photochemical Reactor, model 1061 (total output: 32 watts, of which about 90% was emitted at the 253.7-nm line), until the starting material has disappeared (approximately 2–3 hours). Evaporation of the solvent and purification by preparative TLC using Merck silica gel HF and benzene-ether (1:1) as the solvent system produces tetramethylcaprolactam, R_F 0.46, in approximately 50% yield; m.p. 113.5–114.5° from pentane.

2. Notes

1. A variety of other cyclic oximes will undergo the photo-Beckmann rearrangement, giving both possible Beckmann products in a 1:1 ratio with retention of stereochemistry at the α-carbon in the case of dissymmetric oximes.[3]
2. The main side products are 2,2,6,6-tetramethylcyclohexanone and the product of reductive cleavage, 2,2,6-trimethylheptanamide.
3. The reaction can be monitored by GLC, using 10% carbowax 20 *M* on high-performance chromosorb in (100/120 mesh).

References

1. R. Cornubert, *Bull. Soc. Chim. France*, 541 (1927).
2. F. Bohlmann and K. Kieslich, *Chem. Ber.*, **87**, 1363 (1954).
3. M. Cunningham, L. S. Ng Lim, and G. Just, *Can. J. Chem.*, **49**, 2891 (1971).

* McGill University, Montreal, Canada.
† University of Arkansas, Fayetteville, Arkansas 72701.

TRICYCLO[4.2.2.0^{2,5}]DEC-9-ENE-*exo,endo* 3,4:7,8-TETRACARBOXYLIC DIANHYDRIDE

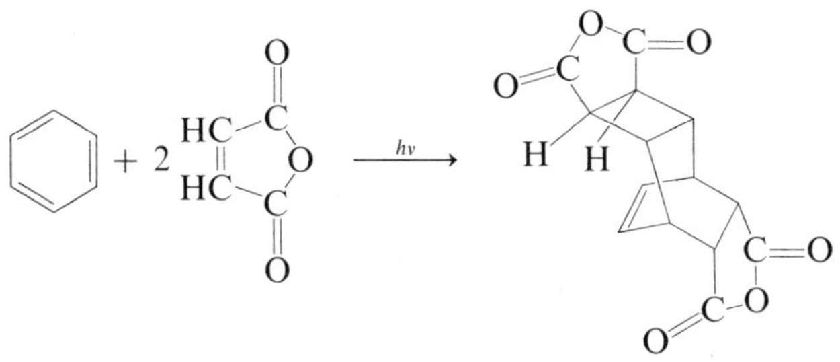

Submitted by ERLING GROVENSTEIN, Jr., DURVASULA V. RAO, and JAMES W. TAYLOR*
Checked by JOHN R. SCHEFFER†

1. Procedure

A solution of maleic anhydride (118 g, 1.20 moles) (Note 1) in acetone (100 ml) (Note 2) and benzene (125 ml, thiophene-free) is placed in a water-cooled quartz cell which encircles a medium-pressure Hanovia lamp (Note 3). The solution after irradiation for 25 hours deposits white crystals on the walls of the cell. These are separated by filtration and washed with acetone to give 10.5 g of adduct (I). Another 6.5 g of adduct (I) may be recovered by evaporation of the solvent, boiling the residue with 200 ml of acetone to dissolve unreacted maleic anhydride, and separation of the adduct by filtration (Note 4). The total yield of adduct (I), m.p. 355–357° (dec), is about 17.0 g (10%) (Note 5).

2. Notes

1. The maleic anhydride was freshly distilled before use.
2. The acetone was a commercial grade which was distilled from potassium permanganate.
3. The lamp was a 1200-watt quartz mercury lamp (type LL) of 30-cm arc length manufactured by Hanovia Lamp Division of Engelhard Industries, Inc. The temperature of the lamp was further regulated by blowing a stream of air (from a small fan) through the annular space between the lamp and the inner cell wall.

* Georgia Institute of Technology, Atlanta, Georgia 30332.
† University of British Columbia, Vancouver, B.C.

4. The proportion of product which separates as crystals during irradiation is somewhat variable; if the solution is allowed to stand 24 hours after completion of the irradiation, only 1 to 2 g of product remains dissolved in the acetone-benzene solution.

5. A period of irradiation of 48 hours in the present apparatus increased the yield of adduct only 1 to 2 g and gave rise to a more colored product and solution.

3. Methods of Preparation

The present procedure is similar to that previously described.[1] An alternative method[2] consists of irradiation of a solution of maleic anhydride in benzene with benzophenone as a photosensitizer (inert atmosphere). If the irradiation of maleic anhydride in benzene is conducted in the presence of 0.825 *M* trifluoroacetic acid, the formation of adduct I is completely suppressed and, in its place, phenylsuccinic anhydride is formed.[3] If the irradiation is conducted in the presence of duroquinone, a 1:1:1 adduct of benzene, maleic anhydride, and duroquinone results.[4] Adducts analogous to I (but with the heterocyclic O replaced by NH or NR) are formed by irradiation of benzene solutions of maleimide and some *N*-substituted maleimides.[5]

References

1. E. Grovenstein, Jr., D. V. Rao, and J. W. Taylor, *J. Am. Chem. Soc.*, **83**, 1705 (1961). See also H. J. F. Angus and D. Bryce-Smith, *J. Chem. Soc.*, 4791 (1960).
2. G. O. Schenck and R. Steinmetz, *Tetrahedron Lett.*, 1 (1960).
3. D. Bryce-Smith, R. Deshpande, A. Gilbert, and J. Grzonka, *Chem. Comm.*, 561 (1970).
4. G. Koltzenburg, P. G. Fuss, S.-P. Mannsfeld, and G. O. Schenck, *Tetrahedron Lett.*, 1861 (1966).
5. D. Bryce-Smith and M. A. Hems, *Tetrahedron Lett.*, 1895 (1966).

2,6-*endo*-TRICYCLO[5.3.1.$0^{2,6}$]-11-OXO-UNDEC-9-ENE

OCH$_3$ O

$\xrightarrow{h\nu}$ $\xrightarrow{H^+}$

OCH$_3$

Submitted by G. SUBRAHMANYAM*
Checked by JOSÉ ORS*

1. Procedure

A solution of anisole (30 ml) and cyclopentene (60 ml) in cyclohexane (30 ml) (Note 1) is irradiated in a cylindrical quartz tube (30 × 8 cm O.D.), closed with a serum cap in a Rayonet RPR-208 reactor fitted with four 253.7-nm lamps (Note 2) for 70 hours. From the yellow reaction mixture solvents are removed in a rotary evaporator (Note 3) and this residue is connected to a vacuum pump (~2 mm) for 15 minutes to remove any volatile material. The pale yellow oily residue containing the adduct and anisole (27.2 g) (Note 4) is refluxed with conc. HCl (10 ml), water (20 ml) and acetone (80 ml) for 2 hours on a water bath (Note 5). Part of the acetone is removed under suction from the resulting green reaction mixture and water (150 ml) is added and extracted with methylene chloride (3 × 50 ml). The extract is washed with saturated sodium bicarbonate solution (3 × 50 ml), then by water (2 × 50 ml), and dried over anhydrous sodium sulfate; the solvent is then removed. The crude product (~22 g) (Note 6) is distilled on a Vigreux column under vacuum (~1 mm). After a forerun of anisole (Note 7), the tricyclic ketone distills at 78–80° as a pale yellow oil, 7.2 g. The product is (95%) pure. The nmr spectrum has been reported.[1]

2. Notes

1. Anisole of 98% purity supplied by Chemical Samples Co., cyclopentene, research grade, supplied by Philips Petroleum Co., and cyclohexane, spectroquality, supplied by Matheson, Coleman and Bell, were used.
2. The reactor, supplied by Southern New England Ultraviolet Co., Middletown, Connecticut, was employed. The unfilled portion of the

* IBM Watson Research Center, Yorktown Heights, New York 10598.

quartz tube was wrapped with aluminum foil to avoid irradiation in vapor phase.

3. The water-bath temperature was kept below 60° during the removal of the solvents.

4. The weight of the crude product is variable, depending on the amount of anisole left behind.

5. Within 10 minutes the reaction mixture assumes a green color which deepens with time.

6. The weight of the crude product varies slightly in each experiment.

7. Most of the early fraction (~5–6 g) distilled below 30° at 2 mm.

3. Merits of the Preparation

The tricyclic ketone is prepared with commercially available starting materials in two simple steps. The procedure can be extended by starting with substituted anisoles and cyclopentenes to obtain synthetically important tricyclic ketones.

Reference

1. R. Srinivasan, V. Y. Merritt, and G. Subrahmanyam, *Tetrahedron Lett.*, 2715 (1974).

2,3,3-TRIMETHYL-2,3-DIHYDRONAPHTHO-[2,1-b]THIOPHENE-1,1-DIOXIDE

Submitted by A. G. SCHULTZ and M. B. DeTAR*
Checked by J. ERHARDT and B. G. KURR*

1. Procedure

A. 2-(β-Naphthyl)isopentenyl sulfide. A solution of β-naphthalenethiol (10.0 gm, 0.0635 mole) (Note 1), 3-methyl-2-butanone (13.0 ml, 0.123 mole) (Note 1), and *p*-toluene sulfonic acid (~50 mg) in benzene (40 ml) is placed in a 200-ml, round-bottomed flask equipped with a magnetic stirring bar, water separator, and condenser. Dry nitrogen is passed into the solution for 10 minutes, after which a nitrogen inlet is positioned atop the condenser. The solution is rapidly refluxed until a calculated amount of water is collected in the water separator (Note 2). The reaction is cooled to room temperature; benzene (50 ml) is added and the solution is washed with 1 *N* sodium carbonate (2 × 10 ml). The benzene layer is dried by suction filtration through a pad of anhydrous magnesium sulfate. Solvent is removed on a rotary evaporator and the remaining oil is transferred to a vacuum distillation apparatus (~10 cm Vigreux column). Distillation gives sufficiently pure 2-(β-naphthyl) isopentenyl sulfide (12.1–12.7 g, 85–90%), b.p. 148–150° (~0.2 mm), which solidifies on standing, m.p. 46–48°.

B. 2,3,3-Trimethyl-2,3-dihydronaphtho[2,1-b]thiophene (Note 3). A solution of 2-(β-naphthyl)isopentenyl sulfide (12.1 g, 0.0531 mole, 0.152 *M*) in spectral grade benzene (Note 1, 350 ml) is purged with argon for 20 minutes and irradiated with a 450-watt Hanovia medium-pressure mercury arc

* Department of Chemistry, Cornell University, Ithaca, New York 14853.

lamp in a Pyrex immersion well as a slow stream of argon is passed into the solution. Reaction progress is conveniently followed by nuclear magnetic resonance spectroscopy; the methyl groups in 2-(β-naphthyl)isopentenyl sulfide appear as a complex multiplet at δ 1.8 to 2.2, whereas 2,3,3-trimethyl-2,3-dihydronaphtho[2.1-b]thiophene gives resonances at δ 1.61 (s), 1.28 (s), and 1.32 (d, J = 7.0 cps). After conversion is judged to be greater than 95%, solvent is removed on a rotary evaporator and the brown residue is transferred to a short-path vacuum distillation apparatus (Note 4). Distillation (Note 5) produces nearly pure 2,3,3-trimethyl-2,3-dihydronaphtho[2,1-b]-thiophene (10.5–10.8 g, 87–90%), b.p. 126° (~0.08 mm).

C. 2,3,3-Trimethyl-2,3-dihydronaphtho[*2,1-b*]*thiophene-1,1-dioxide.* A solution of 2,3,3-trimethyl-2,3-dihydronaphtho[2,1-b]thiophene (0.75 g, 0.0033 mole) in methylene chloride (10 ml) is stirred at room temperature as a solution of *m*-chloroperbenzoic acid (1.50 g, 85% active, ~2.2 equivalents) in ether (10 ml) is quickly added. After 4 hours methylene chloride (50 ml) is added and *m*-chlorobenzoic acid is extracted with 1 *N* sodium carbonate (4 × 20 ml). The methylene chloride layer is dried by suction filtration through a pad of anhydrous magnesium sulfate and solvent is removed on a rotary evaporator. Crystallization of the resulting oil from ether-petroleum ether gives analytically pure 2,3,3-trimethyl-2,3-dihydronaphtho[2,1-b]thiophene-1,1-dioxide (0.75 g, 88%), m.p. 160–162° (Notes 6 and 7).

2. Notes

1. Materials were obtained from Eastman Kodak Co. and used as received.
2. It is essential that oxygen be removed from the reaction medium because thiols undergo a facile oxidative dimerization to form disulfides when heated in the presence of oxygen. The reaction may be monitored by vapor-phase chromatography by using a 6 ft × 1/8 in. column filled with 10% UC-W98 on Chromosorb W 60/80 mesh at 215°; retention time: thiol, ~4 minutes; product, ~15 minutes.
3. For related examples see ref. 1.
4. Approximately 23 hours of irradiation time is required for greater than 98% reaction.
5. The only detectable impurity is 2-(β-naphthyl)isopentenyl sulfide.
6. Proton nmr ($CDCl_3$, tetramethylsilane as internal standard): δ 1.52 (3H, d, J = 7.0 Hz), 1.53 (3H, s), 1.76 (3H, s), 3.34 (1H, q, J = 7.0 Hz) and 7.0 to 8.3 (6H, m); ir (KBr): 7.74 (s) and 8.91 μ(s).

7. This entire sequence has been successfully incorporated into an advanced undergraduate experimental laboratory course at Cornell University.

Reference

1. A. G. Schultz and M. B. DeTar, *J. Am. Chem. Soc.*, **96**, 296 (1974), and A. G. Schultz and M. B. DeTar, *ibid.*, June (1976).

TRIMETHYLPENTACYCLO[5.3.0.0^{2,10}.0^{3,5}.0^{6,8}]DECANE-1,4,7-TRICARBOXYLATE

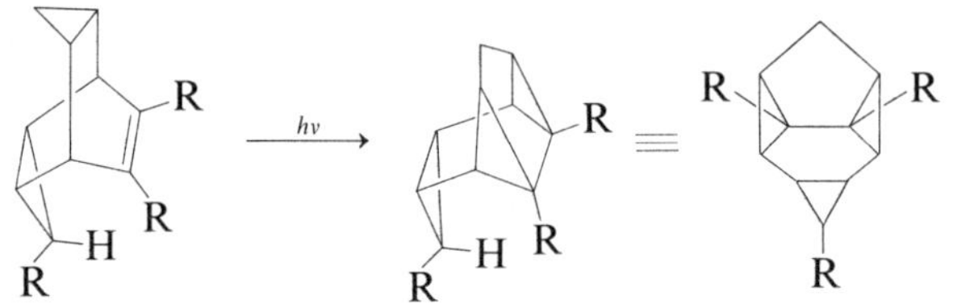

$R = CO_2CH_3$

Submitted by W. EBERBACH and H. PRINZBACH*
Checked by J. CORNELISSE and Mrs. G. M. GORTER-LAROY†

1. Procedure

A solution of trimethyl *exo,endo*-tetracyclo[3.3.2.0^{2,4}.0^{6,8}]dec-9-ene-3,9,10-tricarboxylate (1.0 g, 0.0032 mole) (Note 1) in 280 ml of pure acetonitrile is placed in a cylindrical irradiation vessel equipped with a nitrogen inlet at the bottom and a water-cooled 1-mm-thick Vycor immersion well containing a 70-watt Hanau high-pressure mercury lamp. The reaction vessel is cooled to −30°C and irradiated under a nitrogen atmosphere (Note 2). The reaction is monitored by uv spectroscopy, taking advantage of the decreasing absorption shoulder at 239 nm. After 4 hours the solvent is removed under reduced pressure and the solid residue is recrystallized from *n*-hexane/ethyl acetate (6:1) affording 0.94 g (94%) of photoproduct, m.p. 117–117.5°C (Note 3).

2. Notes

1. *exo*, *endo* with respect to the C═C double bond; it is prepared in 66% yield[1] by copper-catalyzed addition of methyl diazoacetate to dimethyl tricyclo[3.2.2.0^{2,4}]nona-6,8-diene-6,7-dicarboxylate.[2]
2. A temperature increase of −10 to −5°C during the reaction is not critical for the desired transformation.
3. As shown by vpc analysis of the crude material, the only by-product in this reaction is a minor component (1%) of so far unknown structure (*vide infra*).

* Universität Freiburg, D-7800 Freiburg, West Germany.
† Gorlaeus Laboratoria, Rijksuniversiteit, Leiden, The Netherlands.

3. Methods and Merits of the Preparation

The photoisomerisation described here is a typical example of the class of [$2\pi + 2\sigma$]-cycloaddition reactions. It proves to be a rather general and useful synthetic route to highly strained polycyclic systems.[3] This example convincingly shows the importance of the relative arrangement of the three-membered ring and the C=C-double bond on the course of this type of photocycloaddition reaction[4] (the 1% by-product might be the isomer formed by reaction of the C=C-bond with the *endo*-three-membered ring), an observation that is now well documented. This synthesis can be adapted to other derivates of the title system.

References

1. W. Mayer, Staatsarbeit, Universität Freiburg, 1969.
2. M. J. Goldstein and A. H. Gevirtz, *Tetrahedron Lett.*, 4413, 4417 (1965).
3. For leading reference see H. Prinzbach and D. Hunkler, *Chem. Ber.*, **106**, 1804 (1973).
4. H. Prinzbach and W. Eberbach, *Chem. Ber.*, **101**, 4083 (1968); P. K. Freeman, D. G. Kuper and V. N. M. Rao, *Tetrahedron Lett.*, 3301 (1965).

INDEX